Suffering for Science

Suffering for Science

Reason and Sacrifice in Modern America

REBECCA M. HERZIG

RUTGERS UNIVERSITY PRESS

NEW BRUNSWICK, NEW JERSEY, AND LONDON

LIBRARY OF CONGRESS CATALOGING-IN-PUBLICATION DATA

Herzig, Rebecca M.
  Suffering for science : reason and sacrifice in modern America / Rebecca M. Herzig.
    p. cm.
  Includes bibliographical references and index.
    1. Science—Social aspects—United States—History—19th century. 2. United
States—History—19th century.   3. Body, Human—Social aspects—United States—
History—19th century.   4. Self—History—19th century.   I. Title.
  Q127.U6H396 2005
  509'.73'09034—dc22

                                                                        2005002577

British Cataloging-in-Publication information for this book is available
from the British Library.

Manufactured in the United States of America

For
Sidney J. Herzig
(1937–2004)
and
Norman O. Brown
(1913–2002)

CONTENTS

PREFACE

From chemists poisoned by mercury spills to geologists caught in the path of active volcanoes, countless stories of contemporary scientific practice propose that a life in science may be uncommonly painful, if not downright lethal.[1] Most remarkable about these tales of pain is that so many focus not on unwitting accident or misfortune (the primatologist who contracts a rare disease, the oceanographer who boards a doomed research vessel) but on individuals' *willing* embrace of suffering, a deliberate surrender of safety and comfort. To study avalanches, for instance, two civil engineers in Montana repeatedly bury themselves in the rushing tons of snow. "There is a science to all this," one of the researchers tells an incredulous reporter. "I swear."[2] A group of AIDS researchers, declaring that they are not simply suicidal, offer themselves as the first human subjects for an experimental vaccine based on the human immuno-deficiency virus.[3] A molecular biologist insists that even mundane laboratory work necessarily entails the "sacrifice" of "community, family, and self."[4] The author of a recent children's book condenses these themes to a single, stark sentence: "Science is horrible, and just as you suffer in science lessons, so scientists suffer for science." Scientists endure pain and suffering, the book concludes, not "to make life more comfortable for the rest of us" but for one overarching reason: "because they think that science is fascinating."[5]

Notice that in these stories, science—that standard of all things judicious and disinterested—relies on a curious counterpart: the enthralled investigator willing to endure almost any manner of pain. Science proceeds in no small measure due to scientists' devotion—devotion at once reasonable and compulsive, voluntary and involuntary. Something about science "gets under your skin," explains physicist Alan Lightman in a recent edition of the *New York Times,* "keeps you working days and nights at the sacrifice of your sleeping and eating and attention to your family and friends." Scientists "do what they do because they love it, and because they cannot imagine doing anything else. In a sense, this is the real reason a scientist does science. Because the scientist must."[6]

This book is an effort to understand the recurrence of such themes of will, compulsion, and sacrifice in science—a domain of life often depicted as the epitome of secular reason and liberal political thought. While I hope that the book's ruminations on voluntary suffering will hold some relevance for contemporary scientific practitioners (as I reconsider briefly in the epilogue), ultimately this is a study of the past rather than the present. For the question I wish to pose is not "Why does science demand sacrifice?" but "Why do we *say* that science demands sacrifice?" From where did we get the idea that one must suffer in order to "do" science or be a scientist?

In presenting a history of the presumed link between science and suffering, I offer an alternative to those who portray science as a timeless, inhuman force to which we are haplessly subjected. While studying this history surely won't free anyone from the tendency to couple knowledge and pain, it might help arouse further conversation about aspects of daily life generally left unexamined. Rather than insisting that science requires our sacrifice or compels us to suffer, perhaps we lovers of knowledge will begin to hold one another accountable for the arrangements of social life we create and maintain. Of course, some of us may well continue to subject ourselves to suffering, but at least we might begin to reckon with the forms of pleasure we take in doing so.

Suffering for Science

Introduction

Truth at Any Price

In 1802, a medical student from New Jersey named Stubbins Ffirth sought to determine the cause of contagion for yellow fever, a disease then so widespread that it was often referred to as the "American plague." At the time of Ffirth's investigations, prevailing wisdom held that the fever spread through its characteristic "black vomit," a discharge darkened by the victim's internal bleeding. To determine whether the vomit did in fact convey disease, Ffirth collected a large quantity of it from dying fever patients. Then, on the fourth day of October 1802, Ffirth began to experiment: "I made an incision in my left arm, mid way between the elbow and wrist, so as to draw a few drops of blood; into the incision I introduced some fresh black vomit."[1]

After the inflammation stemming from the initial incision subsided, Ffirth obtained some additional "fresh black vomit" and inserted it into his other arm. Once this second incision also healed without incident, Ffirth continued to expose himself to the purported contagion: he dropped black vomit in his eye; he cooked black vomit in a skillet and inhaled its rising steam; he vaporized some black vomit and spent two hours in a small, hot room breathing the resulting fumes; he swallowed pills made from black vomit; and he quaffed a solution of water and recently ejected black vomit. After this series, Ffirth flirted briefly with the idea of "desisting from any further experiments" but ultimately remained steadfast in the variation and repetition of his trials.[2] Upon concluding that black vomit did not cause the fever, the meticulous Ffirth then undertook another set of experiments on his own body, this time with the blood, saliva, perspiration, bile, and urine of yellow fever victims.

Ninety-eight years later, a small group of people undertook a second set of experiments with yellow fever, this time performed under the auspices of the U.S. army surgeon general. The Yellow Fever Commission, also referred to by the name of its head researcher, Walter Reed, was comprised of four physicians:

Reed himself, Aristides Agramonte, James Carroll, and Jesse Lazear. Building on Cuban physician Carlos Juan Finlay's proposal that yellow fever was conveyed by mosquitoes, several members of the surgeon general's commission traveled to Cuba to test the theory. The researchers fed a number of the insects on the blood of yellow fever patients and then captured the infected mosquitoes in glass tubes. Two of the commission members, Jesse Lazear and James Carroll, placed the tubes against their forearms and abdomens, allowing the mosquitoes to feed on their blood. Both Carroll and Lazear quickly fell ill with the fever; and twelve days after the experiment commenced, Jesse Lazear died, delirious and vomiting. After Lazear's death, the team conducted another round of experiments, this time using volunteers from the army medical corps and a number of recent emigrants from Spain. Combining the evidence from all of the experiments, the Reed Commission proclaimed the mosquito, an intermediate host, to be the agent of yellow fever transmission.[3]

At first glance, the experiments conducted by the Reed Commission mirror those performed a century earlier. Like the investigators in Cuba, Ffirth sought to solve the riddle of yellow fever, was committed to experimentation in the resolution of that problem, considered his own body an appropriate experimental tool, and publicized the results of his investigations. Yet the surface similarities between the two episodes quickly dissolve to reveal their more profound differences. Quite unlike Ffirth, the men involved in the Reed experiments were hailed in their own time as noble sufferers, willing to sacrifice their lives for science. Reed, although himself not actually on the island at the time of the historic self-infections, was soon the beloved subject of dozens of sculptures, paintings, and articles.[4] He took honorary degrees from Harvard and the University of Michigan and ultimately became the namesake of the most prestigious army hospital in the United States, wherein American presidents often still receive medical treatment. The experiments themselves were later recounted in a number of popular forms, including Sidney Howard's hugely successful play, *Yellow Jack,* and a lucrative Hollywood film of the same name.[5] To the question, Is it worth a human life to find the cause of yellow fever? one best-selling 1926 book declared that members of the Reed Commission boldly answered yes.[6]

Narratives of self-sacrifice were not limited to popular representations of the Reed experiments. Descriptions of noble martyrdom were equally prevalent in the original reports of the experimental results in professional journals, congressional records, and international scientific meetings. Surgeon General George Sternberg described Lazear as a "highly esteemed" gentleman who died "a martyr to his scientific experiment."[7] Writing in the journal *Medical Record* in 1901, head researcher Reed similarly stressed the special moral character of the experimenters. His fallen colleague, Reed wrote, displayed "manly and fearless devotion to duty such as I have never seen equaled. In the discharge of [his duty], Dr. Lazear seemed absolutely tireless and quite oblivious of self. Filled with an earnest enthusiasm for the advancement of his profession and for the cause of science, he let no opportunity pass unimproved."[8] John Kissinger and John Moran, two Americans who volunteered for further experiments after Lazear's gruesome death, were also alleged to have done so "solely for the cause of humanity and in the interest of science." The men reportedly accepted the life-threatening task only on the condition that they receive no financial compensation for it.[9]

In contrast, the "cause of science" that figures so prominently in Walter Reed's account is entirely absent in Ffirth's report. Ffirth neither presented his actions as demanded by science nor used the language of manly sacrifice when reporting his findings. That Ffirth repeatedly attempted to infect himself with an incurable and deadly disease merited no special comment in his treatise on fever.[10] His only articulated hopes for the experiment were (1) the preservation of human life and (2) the revision of ill-conceived quarantine laws.[11] While Ffirth did remark on the "valour of Fredonia's sons" who fought for colonial independence, the rhetoric of soldierly mettle never bled into his experimental undertakings.[12] In short, no matter what sorts of things Ffirth might have been doing with black vomit, he never hinted that his actions were undertaken on behalf of science. Less than a century later, however, the significance of the scientist's voluntary suffering was readily apparent, if not fully understood. As one periodical responding to the deaths of Lazear and other experimenters noted in 1901, such expenditure for the "sake of science" represented a "form of self-sacrifice with which we are not familiar, and to which we have not yet adjusted ourselves."[13]

Emerging in the space between these two trials is the possibility of imagining science as an entity worthy of self-sacrifice and the concurrent possibility of imagining scientists as those unusual persons willing to suffer and die in its name. That these possibilities were not yet taken for granted—that Americans had not yet "adjusted themselves" to suffering for science—is evident in the outpouring of statements on sacrifice by and about American scientists that began after 1875. "Higher than all," declared the editors of the new national journal *Science* in 1883, science "must be devoted to the truth. It must cheerfully undertake the severest labor to secure it, and must deem no sacrifice too great in order to preserve it."[14] Mathematician George Bruce Halsted echoed this sentiment in 1896, urging investigators to "sacrifice all unflinchingly" for science, "the benign empress of our modern world."[15] Physicist Michael Pupin carried the theme into the twentieth century, affirming that a life of science "cannot be attained without unceasing nursing of the spirit and unrelenting suppression of the flesh."[16] From diverse disciplinary, institutional, and regional locations, such commentators called on individual investigators to surrender physical comfort, material gain, and social solace for the sake of science. The writings they generated reveal not only an important presumption about the relative value of scientific knowledge (as one chemist put it in 1895, "a human life is nothing compared with a new fact in science") but also two far more critical claims: that the advancement of science *requires* painful self-sacrifice and that scientists are uniquely willing—even eager—to take on this pain.[17]

This book follows these twined themes, elucidating the conditions in which late nineteenth-century Americans came to characterize science as an autonomous and exacting entity and scientists as those subjects specially beholden to its costly demands. It shows how these two mutually constitutive figures—masterful science and submissive scientist—were enabled and sustained by the pliant ligature binding them together: self-sacrifice.

The very word *sacrifice* displays this pliancy. Derived from the Latin *sacer,* it wobbles between the contrary roots "hallowed" and "detestable."[18] While retaining the elasticity of its classical cousin, the decidedly more recent term *self-sacrifice* (the *Oxford English Dictionary* lists William Wordsworth's 1805 "Ode to Duty" as its earliest use) introduces all the further equivocation contained in modern concepts of self. Confusingly, the self is at once subject and object of the self-sacrifice, the agent determining action and the substance acted upon. The layered connotations of self-sacrifice help explain its prominence in writings of the late nineteenth century, when received relationships among selfhood, personhood, property, and embodiment were in the process of being upended and reworked. Wherever self-sacrifice flourished, we find deliberation on these relationships: scrutiny of the nature of free will, the purpose of suffering, and the limits of reason. The book's central concern, in the end, is with such deliberations—that is to say, with changing constitutions of the human.

Through this examination of bodies, selves, and persons, I revisit several of the central issues of the Gilded Age and Progressive Era, including relations between religiosity and secularism, tensions between industrialization and democratization, and connections between racial and sexual exclusion and liberal political traditions. With respect to these wider subjects, the book presents three interwoven claims. First, I show how self-sacrifice for science, an ethic forged by the larger upheavals of Civil War and Reconstruction, presupposed a distinctly privileged kind of self, one characterized by specific histories of property, personhood, and civic participation. Self-sacrifice implied a willing surrender of one's person (that is, one's property); but only some bodies were historically endowed with the kind of self-ownership requisite to such willful, free forfeiture. Second, I emphasize the paradoxical quality of that social privilege: a power made evident through scientists' sunken eyes, emaciated bellies, and bloody stumps. The blurring of power and vulnerability, elevation and humiliation, and pleasure and pain evident in sacrifice for science reflects not only the abiding influence of Protestant doctrines of salvation but also the received contradictions of the modern liberal subject. Third, I demonstrate that the ambiguities inherent to voluntary suffering spawned a set of persistent questions for late nineteenth-century scientists: just what might be gained by forgoing sleep, skipping meals, or forfeiting limbs? Were such "sacrifices" relinquishments made with no expectation of return, or were they acts of calculated reciprocity—bartered for some new reward of equal or greater value? As will be seen, commentators' discordant responses to these questions point to the consequential new role then being allotted to science: the task of addressing and answering perennial problems of ultimate meaning and purpose.

These interwoven threads become more readily apparent if we return to the Reed yellow fever experiments. Given continued U.S. military presence in Cuba after the Spanish-American War, the freedom of individuals on the island was in question in the American, Cuban, and Spanish presses even before the famous trials. As historian Susan Lederer reports, to quell concern over the free will of volunteers in the experiments (particularly the independence of the recent emigrants from Spain), Reed took the then unusual step of requiring individual written consent forms and restricting participation to those over the age of twenty-four, the age of consent in Spain.[19] When publishing their experimental findings, Reed and his colleagues were similarly careful, stressing that "all experiments were performed upon persons who had given their free consent."[20]

Despite Reed's careful efforts, free consent remained a slippery category: only some bodies were apprehended as fully consensual and hence as authentically sacrificial. Participants fell beneath the standard of sacrificial subjectivity by contractually consigning themselves for money—that is, by bartering their bodies rather than offering them without expectation of return. The receipt of payment, one hundred dollars in gold, suggested obligation to the payer and thus troubled the understandings of free gift (and autonomous selfhood) inherent to the concept of self-sacrifice. As a result, the newly arrived Spanish emigrants, who accepted gold for their participation in the experiments, generally were unacknowledged in the numerous memorial tributes to the "martyrs" of the experiment constructed at the time. Other tangible forms of memory—monetary awards, plaques, statues, medals, honorary degrees, textbooks—reproduced this exclusion, highlighting the "manly devotion" of Carroll, Lazear, Kissinger, and Moran and ignoring the illnesses and deaths of others.[21] Similar patterns of differentiation appear in each of the examples discussed in the book. For one's action to be self-sacrificial, it must be freely chosen. Freedom, however, was always malleable and situation: a contingent, relational attribute determined by specific historical conditions.

These conditions were in flux after 1875 as the United States concluded the lengthy, bitterly contested transition from slavery to free labor. As previously dispossessed subjects came to own their own bodies, to be "persons" in that liberal sense, they acquired the freedom to sell or rent their bodily capacities. The elevation of market exchange as the model of all relations among persons thus thrust to the fore the nature of consent, the principle of individual liberty upon which such relations of exchange were based. In this setting, as historian Amy

Dru Stanley has shown, the contract became at once a material transaction (an identifiable set of legal and economic associations) and a figurative tool (a generalized vocabulary for framing the relationship between self and other). Paradoxically, the growing importance of consensual contract, which determined individual volition to be the defining element of the self, highlighted the obligations and imperatives of life under industrial capitalism.[22] Scientists joined other late nineteenth-century Americans in coming to terms with renewed forms of coercion and endeavoring to delineate some fundamentally inalienable, free self. Like labor reformers, freed slaves, feminists, and bonded immigrants, scientists struggled to live with the dilemmas of consent and compulsion, liberty and obligation effected by the abolition of slavery and the concurrent expansion of industrial capitalism.

Self-sacrifice raised the specter of obligatory exchange in an especially acuminate way. The phrase *self-sacrifice* implies a loss suffered, an uncompensated expenditure, an unrestricted offering. A sacrifice, *Webster's* summarizes handily, is "something given up." For scientists as well as for artists, theologians, military officers, and the other late nineteenth-century Americans enamored of the phrase, the normative power of self-sacrifice lay precisely in this meaning: in its confounding of the usual logic of consensual exchange. As one proponent put it, "a sacrifice which involves no real loss and impoverishment, or which ultimately leads to personal gain, is not true self-sacrifice."[23] To remain outside the bounds of ordinary transactions in the marketplace, the offering of self must not be restituted. The calculated sacrifice ("is it worth a human life to find the cause of yellow fever?"), a gift tendered with expectation of return, implies no sacrifice at all. At stake in scientists' calls for self-sacrifice was thus not merely the value of the parts of the self surrendered as compared to the forms of knowledge obtained: whether, say, some new fact was "worth" a whole finger or just the tip. More important, the matter at hand was whether science might present an alternative to the tyrannical logic of the marketplace—whether it could, or ought to, be freed from the profane presumptions of the self-interested contractual exchange.

Such dilemmas were the subject of helpfully explicit debate among the scientists described in subsequent chapters: university-based researchers in the 1870s and 1880s, polar explorers in the 1890s and 1900s, X-ray experimenters in the 1900s and 1910s, and the fictional bacteriologists depicted in Sinclair Lewis's 1925 *Arrowsmith*. Discussions of will, reason, and sacrifice recur across their diaries and letters, funeral orations and inaugural addresses, laboratory notebooks, and published reports. That many of these writings were ritualistic in nature does not dampen their utility as sources for historical analysis, for even their rhetorical excesses suggest the significance of governing norms and ideals. Drawing on these sources, I map the place of self-sacrifice during a pivotal moment in the history of American science. From the 1870s to the 1920s and beyond, a malleable ethic of voluntary suffering helped unite practitioners marked by increasing institutional diffusion and specialization as members of the imagined, undying body of science.[24]

To be sure, the cases discussed in this book are not comprehensive. Most of the "Americans" of whom this study speaks spent their lives in northern states, many in New England.[25] While their positions afforded them the ability to present their local experiences as those of the nation as a whole, it is worth remembering that their assertions about America were, like any others, always partial. When I describe the claims of a few investigators as indicative of larger national trends, I do so because to some degree these investigators were correct in sensing themselves to be at the center of American scientific work. By 1845, Boston surpassed all other U.S. cities in numbers of scientists in residence (despite having merely a fraction of the population of Philadelphia or New York), and the region's influence over the direction of the sciences in the United States persisted into the twentieth century. Although this handful of elite New England scientists was by no means wholly representative of all inquiry ongoing within U.S. borders, scientific centers held disproportionate influence over the shape and character of research in the peripheries, as they continue to do so today.[26]

Similarly, as much as the examples I selected stress the prominence of self-sacrifice in the five decades after Reconstruction, voluntary suffering was by no means the only standard of scientific progress available in the late nineteenth century. Some commentators emphasized the importance of the fortuitous scientific accident, as in Antoine-Henri Becquerel's discovery of natural radioactivity (said to have resulted from leaving some uranium rocks in a drawer containing photographic plates) or Luigi Galvani's theory of "animal electricity" (said to have been detected when a charged metal scalpel happened to contact the bared nerve of a dissected frog). Others emphasized the pleasure and merriment of scientific pursuits: ethnologist Steward Culin (1858–1929), for example, often underscored the importance of childish play in the generation of new knowledge. Serendipity and ease vied with voluntary suffering in numerous late nineteenth-century accounts of scientific practice. While there was never universal accession to calls for suffering, self-sacrifice did prove consequential during a period of profound transformation in the shape and scope of scientific labor. This book seeks to explain why, given the existence of other, less painful alternatives, so many scientists chose to align themselves with this ethos and considers, however speculatively, some of the lasting ramifications of this decision.[27]

OF COURSE, nineteenth-century American scientists were not the first to affiliate knowledge with pain. Varied experiences of suffering have long pervaded representations of the contemplative life. In the dominant traditions of western intellectual history, the passage from ignorance to enlightenment has been portrayed as ineluctably arduous and bloody, if not downright lethal. From Plato's den-dwellers scraping their way out of the allegorical cave to the *Agamemnon*'s claim that "wisdom comes alone through suffering" (*pathos mathei*), classical Greeks crafted particular associations between truth and pain.[28] Suffering's value as a sign of chosenness, as an elect point of access to holy wisdom, has also been central to Jewish philosophical traditions, evident not only in the Hebrew Bible but also ancient and medieval rabbinic literature.[29] Early Christians, in turn, developed their own gruesome customs of suffering—drinking pus, whipping themselves with nettles, and so forth.[30] Even today, the age-old affinities between truth and suffering persist in mundane ways, coagulating around English words such as *painstaking* and *labor*. To "take pains" with a task is to move it closer to perfection.[31]

Despite the obvious longevity of the affiliation between truth and pain, the striking recurrence of self-sacrifice in late nineteenth-century science begs further explanation. For the persistence of voluntary suffering defies the common belief that detachment from such customs is precisely what distinguishes science from other forms of knowledge—that science is, in anthropologist Sharon Traweek's phrase, a "culture of no culture."[32] A well-established body of scholarly literature arises from just such a supposition, positing a crucial break in the early modern period. At this time, the literature suggests, an "invisible, autonomous, virtual" observer displaced the embodied ascetics of previous knowledge traditions.[33] Where classical Greeks or medieval Christians might have needed to suffer in order to access truth, the modern scientific knower is said to be freed from such requirements—indeed, freed from bodily specificity and individuality altogether. Michel Foucault promoted this view in an essay written shortly before his death:

> In European culture up to the sixteenth century, the problem remains: What is the work I must effect upon myself so as to be capable and worthy of acceding to the truth? To put it another way . . . asceticism and access to truth are always more or less obscurely linked.
>
> Descartes, I think, broke with this when he said, "To accede to truth, it suffices that I be *any* subject that can see what is evident." . . . After Descartes, we have a nonascetic subject of knowledge. This change makes possible the institutionalization of modern science.[34]

In other words, modern science is distinguished by the fact that one no longer need be among the elect—divinely chosen, morally superior—in order to generate knowledge. With the institutionalization of modern science we acquire, perhaps for the first time, the potential to access truth *without* suffering. We have now the "nonascetic subject" of knowledge: any subject can see what is evident.

Given the claim that the institutionalization of modern science enabled the separation of knowledge from the character of the one generating that knowledge, how are we to make sense of the enduring presence of the suffering subject, to understand the tenacity of late nineteenth-century scientists' cries for voluntary sacrifice?

The answer to this question may be found, in part, in the work of those historians, sociologists, and philosophers of science who, in opposition to Foucault, emphasize the ongoing importance of practices of self-constitution in the production of knowledge.[35] Their studies suggest that an ascetic care of the self, including the rigorous regulation of sleep, exercise, diet, and sexual behavior, remained vital to the generation and guarantee of reliable knowledge long after the "break" allegedly marked by Descartes.[36] The contributions of historians Lorraine Daston, Peter Galison, and George Levine have been especially crucial in this respect. Like Foucault, these scholars detect a consequential shift in the nature of the knowing subject, but they place this turn in the nineteenth century, in the shift from Enlightenment ideals of truth to modern notions of objectivity. So divergent were these two positions, Daston proposes, that "proponents of objectivity were sometimes willing to sacrifice accuracy or even truth for its sake."[37]

Even as objectivity marked a break with traditional ways of knowing, however, it maintained some links to earlier practices of the self. An idealized "view from nowhere" presumed a new kind of individual, the subjective self, to which objectivity was opposed. Investigators were encouraged to bury this subjective identity in the "impersonal collectivity" of science, an epistemological ideal which Levine aptly terms "dying to know."[38] Put differently, we might say that care of the self remained fundamental to the generation of knowledge even after the shift from truth to objectivity in the nineteenth century, after the institutionalization of science described by Foucault; but such care became a practice of self-formation in the service of a quite different end. In the age of objectivity, the self is constituted in order to be willfully forfeited.[39]

Given the relevance of "dying to know" to modern notions of objectivity, and given the centrality of objectivity to transnational practices of scientific investigation in the nineteenth century, it is not surprising to find references to suffering among scientists in other national contexts. From the tribulations of Alexander von Humboldt to the wounds of Marie Curie, examples abound of investigators who suffered, often highly visibly, in other settings of modern science.[40] Consider the Duke of Sussex's praise for renowned English astronomer John Herschel at one 1838 dinner. As the large crowd bellowed its approval ("hear, hear"), the speaker recounted how Herschel's strenuous research had been surmounted only by that of a colleague who "died a few years ago a victim to his arduous exertions, in the study of astronomy."[41] Or consider the influential experiments in nerve regeneration conducted by Henry Head and W.H.R. Rivers at St. John's College. After Head had most of the nerves in his left hand and forearm surgically severed, he and Rivers began a series of tests on the sensitivity of the rehabilitating nerves. From 25 April 1903 (the date of the original operation) until 13 December 1907, the two Fellows of the Royal Society met regularly in Rivers's college chamber to apply varying amounts of heat, cold, and pressure to Head's mangled arm and the shaft of his penis. "When the needle was brought into contact with the skin, such as that of the body of the penis, H. was at once conscious that he was being touched with a pointed object. . . . if a sensitive spot had been chosen, H. cried out and started away."[42] As these and myriad other examples attest, American scientists were hardly exceptional in their promotion of volitional suffering; and subsequent chapters recount some of the more important ways in which they adapted norms and practices of sacrifice from other national contexts.

The endurance of the suffering subject in American science, however, is only partly explained by the advent of transnational ideals of objectivity. An equally crucial explanation for scientists' cries for voluntary sacrifice can be found in developments specific to their location. Just as American scientists were not alone among scientists in stressing the necessity and value of voluntary suffering, neither were scientists the only proponents of voluntary suffering to be found in the United States. As this book demonstrates, scientists were part of a much broader invigoration of self-sacrifice during the last quarter of the nineteenth century. In the aftermath of the Civil War and Reconstruction, several trends converged to transform attitudes about suffering held by the well-to-do, including a long tradition of Protestant martyrology, the diffusion of Darwinism, and the steady spread of industrial capitalism. The educated and affluent came not only to naturalize the connection between suffering and advancement but also to glorify it. Voluntary suffering appeared as a cleansing purgative for the nation's gluttonous economic life, a way to renew the moral solidity of a bygone era. By 1870, few men and women of letters blanched when Henry Ward Beecher declared suffering to be the universal "measure of value."[43] Against a backdrop of violent labor strikes, the widespread lynching of African Americans, and continued bloodshed over Indian lands, the pampered began to clamor for pain. Accordingly, a number of white middle-class activities were redescribed in lavishly sacrificial terms. As the ideal investigator was newly lauded for sacrificing himself for science, so, too, the ideal citizen was presumed to offer his life for his country, the artist his life for his art, and the Christian his life (or, less frequently, her life) for God.[44]

I have focused this study of voluntary suffering on American scientists (rather than, say, American mothers) due to the unusually influential character of scientists' engagement with self-sacrifice. Unlike self-sacrificing artists, soldiers, or mothers, scientists both claimed *to* suffer and made authoritative claims *about* the definition of suffering itself: which acts of suffering were normal and which were deviant, which were natural and which were unnatural. Scientists' efforts to document the nature of suffering are evident in numerous examples: in their increasingly elaborate racial taxonomies of pain, their assessments of appropriate and inappropriate uses of anesthesia, and their novel diagnoses of masochism (circa 1886), algolagnia (circa 1901), and other so-called perversions.[45] Another crucial index of late nineteenth-century scientists' rising ability to define and certify suffering is the emergence of sacrifice itself as a distinct object of scientific inquiry. In the massive outpouring of studies such as William Robertson Smith's *Lectures on the Religion of the Semites* (1889), Henri Hubert and Marcel Mauss's *Essai sur la nature et le fonction du sacrifice* (1899), Sir James George Frazer's *The Golden Bough* (1890), Emile Durkheim's *Les formes élémentaires de la vie religieuse* (1915), and William James's *The Varieties of Religious Experience* (1902), writers drew on the increasing legitimacy of science to materialize sacrifice as an empirically demonstrable phenomenon.[46] Quite unlike the self-sacrificing businessman or soldier, the self-immolating scientist could claim to unveil the nature of suffering itself, to reveal "the voice of God . . . in facts."[47] In this respect, late nineteenth-century scientists would soon stand on par with clergy in their ability to influence broader understandings of the nature of suffering.

Scientists' own practices of voluntary suffering—interminable hours spent gathering specimens, painful habits of self-experimentation, perilous ethnological observations, and so forth—were often the very means by which they generated new understandings of pain and volition. Even as scientists were arrogating the authority to sort truly deliberate, self-chosen suffering from mere masochism or barbarism, they themselves embodied such distinctions. They made statements about suffering's true nature even as they enacted, through the medium of their bodies, broader assumptions about the value of suffering. Few other sufferers played such a complicated role, at once trumpeting their own voluntary suffering and reshaping knowledge about suffering itself.

Given the unusual position in which scientists found themselves, maintaining the *reason* for and of one's suffering—establishing that it was freely and purposefully chosen—was therefore of paramount importance, lest the scientist's decision to pass a fourth consecutive evening without sleep or to deliberately contract a deadly disease appear demented. And as one of the central strands running through this book seeks to make clear, the boundaries of reason, the lines between sacrifice and barbarism or pathology, were reliant on other forms of stratification: heroic scientist versus inconsequential indigenous assistant, willing martyr versus perverse masochist. The ability to be a sacrificial self, in other words, was always structured by the ability to consent, itself dependent one one's embodied "location in the material world."[48] Possession of an intentional, free self, a prerequisite for self-sacrifice, was a quality defined only in relation to bodies said to be lacking intention and freedom—in contrast to those bodies in which, as Toni Morrison writes, "the self that was no self made its home."[49] While suffering might, as William James put it, "redeem life from flat degeneration," it could do so only for those historically endowed with the self that *was*.[50]

When explicating varied relations between selves, bodies, and persons in the pages that follow, I employ a range of terms once used to attribute specific characteristics to individuals: *scientist, man, woman, Negro, Chinese, American, slave, citizen, civilized, pathological,* and so on. I forgo incessant quotation marks around such words. I hope it will remain evident throughout that my aim is not to reproduce these categories but to describe the processes by which they were brought into being, contested, and inhabited. The point is not simply to show how different bodies were imbued with different meanings but to illuminate how these bodies came to be differentiated in the first place.[51] To this end, I draw on recent critical historical scholarship on race and sex, which reveals how renewed emphasis on fasting, self-flagellation, and other forms of voluntary suffering in the last quarter of the nineteenth century helped assuage concerns about the changing status of the white American man.[52] To understand self-sacrifice, then, we will first explore the changing contours of freedom and self-possession in nineteenth-century America, particularly their dependence on records of coercion and dispossession.

While I attend to the mutually constitutive relationship between scientists and nonscientists throughout the book, I want to emphasize my abiding focus on the figure of the self-sacrificing scientist, whose paradoxes form the second strand of this study. There are, of course, countless others involved in scientific investigation, including laboratory animals, wives and children, technicians, and other experimental subjects. These others, already skillfully addressed in previous studies, have their own distinctive relationships to the imagined body of science and their own tales of suffering and sacrifice.[53] While the book does address the ways in which practices of self-formation were entangled in practices of domination (for, in the words of literary theorist David Savran, "a penchant for pain by no means rules out the possibility of turning violence against others"), the voluntarily suffering self remains my primary concern.[54] This book is intended to call attention to the forms of domination that elite investigators exercised upon themselves.[55] Attention to these forms of domination reveals a scientific self bristling with uncomfortable contradictions—at once robust and mangled, potent and vulnerable. Following Jesse Lazear and his contemporaries, we are introduced to a subject simultaneously hallowed and detestable, for whom progress occurs only through suffering and liberty requires continuous displays of bondage.

A certain empathy with this ambivalent self prompts the third and final strand of this history of self-sacrifice, a query that initially directed this study: given the presumption that advancement necessitates personal suffering, how did these individuals decide how much of themselves to give? On what grounds were such difficult decisions made? To what end, for what reasons, did these investigators choose to suffer for knowledge?

As subsequent chapters make clear, such questions of reason and purpose were the subject of helpfully explicit debate in the late nineteenth century. In some settings, sacrifice implied calculated exchange—suffering endured with the expectation of the return of equal or greater value. For these advocates, sacrifice appeared eminently generative, destined to return advantageous salaries, professional renown, and useful knowledge to those who weighed their decisions judiciously and wagered appropriately. The American delegates to an 1883 international meeting on the proposed standardization of time and longitude, for instance, reported that the "scientific and practical utility" of joining the international standard "far outweighs the sacrifice of labor and the difficulties of re-arrangement which it would entail."[56] Sacrifice remained an inescapable component of scientific advancement in these portrayals, but proponents framed the relationship as founded on self-evident principles of exchange. The relevant question for these commentators was simply one of predictive accounting: in what sense expenditures were balanced by expected returns. The "only thing to be certain of," one writer reflected in the wake of the lethal Reed experiments with yellow fever, "is that the knowledge gained is worth the possible sacrifice of human life."[57] Again, sacrifice remained firmly within the logic of contract; all that was needed was a clear sense of the price of the capacities surrendered—that is to say, of the relative value of human life.

In other settings, however, the relationship between scientist and science appeared to exceed such reasoned calculation. In these instances, sacrifice signaled divestiture, offerings made expressly because they carried no assurance of return. "Everyone has heard of a meeting of 'men of science,'" one writer recalled in 1921, who toast science "with the acclamation, 'May it never be of any use to anyone.'"[58] Those scientists seeking to counteract the siren voices of materialism, commercialism, and utilitarianism elevated sacrifice precisely because it denied the logic of compensatory exchange.[59] It was the nonreciprocal character of the scientist's labors that assured his separation from the degradation of the modern marketplace. For such advocates, the very meaning of the scientific endeavor lay in the fact that the relation between scientist and science was one of gift rather than contract.

An influential chemist stressed the widespread misunderstanding of this point in 1895: "the most curious misconception is that . . . the aim of science is the cure of disease, the saving of human life. Quite the contrary, the aim of science is the advancement of human knowledge at any sacrifice of human life."[60] For other period commentators featured in this book, sacrifice similarly evokes something closer to orgiastic potlatch than to temperate, planned dispensation. Wounds and privations are treated not as part of a balanced, reciprocal system of exchange but as demonstrations of forfeiture, as acts *not* repaid in kind.[61] A few commentators argued that any effort to put a price to the scientist's suffering was inherently misguided since the exchange could never be assured; after all, one wrote, "only God knows what a fact's worth."[62] The evident lack of economic calculation in the scientist's labor released him from the stench of the contract.

It may be tempting to contend that such references to uncompensated expenditure should not be taken at face value, to insist that scientists' alleged self-sacrifices actually resulted in considerable individual and professional gain. This interpretive stance would hold that whether consciously calculated or not, those who declared disdain for pecuniary or intellectual reward often reaped the tangible benefits of that posture. Some explorers, for example, invoked the romanticism of the commercially futile "polar quest" to hawk everything from motor oil to Bibles; some investigators maintained profitable consulting contracts even while extolling ceaseless devotion to "pure science." Professed aversion to utility and compensation, one might argue, could prove extremely useful—generating increased publication, professional influence, patentable innovations, institutional mobility, and other forms of symbolic capital or financial resources.

However tempting it may be, I invite readers to join me in resisting such an interpretation. While I do note discrepancies evident in various accounts of self-sacrifice, I do not endeavor to root out the eventual personal or collective gain lurking beneath references to deliberate loss. To approach this historical study seeking only to learn how scientists profited from seeming acts of forfeiture would be to miss a subtler and far more interesting point: what was to be gained through voluntary suffering was exactly what was in dispute at the time. From the perspective of the individual trying to decide whether to labor one more hour without food, to march one more mile out into the field, or to run one more experiment before retiring for the night, the outcome of these activities could never be certain, nor could the sensibility of the decision to engage in them. Indeed, it was precisely the enigmatic character of such decisions—of what might be achieved through one's action upon the self—that made the ethic of sacrifice so labile and hence so socially potent.[63]

I use the word *ethic* in this context (rather than *concept, trope,* or *discourse*) for two reasons. First, an ethical frame helps circumvent a simple division between the rhetoric of sacrifice and the "real" experience of suffering it might be said to conceal.[64] Of course, a single act (such as deliberately exposing oneself to yellow fever) might be interpreted by others in qualitatively different ways: as the Ffirth and Reed experiments show, one act might be seen as so routine as to hardly bear mention, while another is rewarded as noble martyrdom, while a third is castigated as gold-grubbing insensibility.[65] I here seek to distinguish between such interpretations while remaining attentive to the specific, concrete practices that render them consequential. Frostbitten toes, distended stomachs, and ulcerated fingers became self-sacrifice (as opposed to accidents or pathologies) only once claimed as deliberate, voluntary contributions to science's ceaseless advancement—and only then when claimed by a particularly located self. Second, an ethical frame helps to maintain a delicate balance between the individual and the collective, an apprehension of the role of social norms in organizing and inspiring action that does not neglect the significance of the acting person. As Stephen J. Collier and Andrew Lakoff explain, problems of ethics—of how one should live—necessarily involve "a certain idea of practice ('how'), a notion of the subject of ethical reflection ('one'), and questions of norms or values ('should') related to a certain form of life in a given domain of living."[66] Approaching self-sacrifice as one such "problem of ethics," this book seeks to understand the interplay of practices, subjects, and norms in the domain of late nineteenth-century American science.

Perceptive period observers on both sides of the Atlantic recognized the fundamental problems of ethics posed by the practice of science. As Max Weber posited in his famous 1918 speech, "Science as a Vocation," whether "what is yielded by scientific work . . . is 'worth being known'" can never be ascertained "by scientific means." The value of the scientist's offering can "only be *interpreted* with reference to its ultimate meaning, which we must reject or accept according to our ultimate position towards life."[67] Other commentators of the era similarly suggested that participation in science depended on some prior faith in its purpose, one every bit as unsubstantiated as belief in God's good will.[68] Ultimately, William James observed in an address to Harvard Divinity students in the fall of 1884, even the scientist's efforts must be remanded to the realm of unreasonable devotion. The scientist's assumption that truth is worth pursuing is "as much an altar to an unknown god as the one that Saint Paul found at Athens. All our scientific and philosophic ideals are altars to unknown gods."[69] Reaching similar conclusions in 1887, Nietzsche emphasized the violent excesses of this faith. The conviction that "truth is more important than any other thing," the "unconditional faith" on which science rests, cannot arise from a reasoned "calculus of utility." Indeed, this faith "must have originated *in spite of* the fact that the disutility and dangerousness of 'the will to truth,' of 'truth at any price' is proved to it constantly. 'At any price': how well we understand these words once we have offered and slaughtered one faith after another on this altar!"[70]

At stake in such provocative reflections, as in our own contemporary assessments of the meaning of sacrifice, is nothing less than the status of the human: the nature of individual freedom, the purpose of suffering, and the possibilities for real and lasting progress. For the scientists discussed here, such enduring questions were the matter of everyday practice as well as grandiloquent philosophical reflection, embodied not only in inaugural addresses and funereal orations but also in each skipped meal and lost hour of sleep. With the purpose (or purposelessness) of voluntary suffering thrown open by the massive social upheavals of the late nineteenth century, the "price" of truth became a subject of renewed concern. In an era of contractual exchange, the power of science to create value was a matter of continual negotiation.

While exploring these negotiations, I underscore the lasting influence of the nation's histories of bondage and disenfranchisement, which gave conviction in "truth at any price" distinctive shape in the late nineteenth century. As I explain in chapter 3, for instance, the "pure scientists" who spearheaded the organization of American research universities defined their lives of toil, loneliness, and penury in ways that excluded the participation of others deemed less worthy of such lofty pursuits. The frozen noses and aching feet of the polar explorers discussed in chapter 4 were bound to histories of selfhood and personhood that endowed some bodies with powers of vulnerability largely denied to others. The few dozen X-ray experimenters described in chapter 5 were hailed as heroic martyrs to science, while the hundreds of laborers and animals subjected to equivalent amounts of X radiation were not. Even the fictional proponents of self-sacrifice considered in chapter 6 reproduce and promote these patterns of differentiation and exclusion. To consecrate oneself to science in the late nineteenth century required a certain kind of socially constituted self. Without this willful self, one could hardly indulge the odd privilege of voluntary suffering.

To restate, the aims of this study of self-sacrifice in American science are threefold: to demonstrate the centrality of self-possession in delimiting which selves could be apprehended as truly sacrificial, to elucidate the paradoxical character of the proprietal self's willful suffering, and to articulate the unsettled character of the reason for (and of) the scientist's sacrificial exchange. Chapters 1 and 2 help frame these arguments: the first by offering a broader history of the voluntarily suffering self, the second by explaining the location of this self in the massive reorganization of scientific labor ongoing after the Civil War. Chapters 3 through 6 flesh out the operations of science, suffering, and selfhood in particular communities of practice. The book's epilogue, "The Ends of Sacrifice," considers the contemporary resonance of these painful histories.

I

Willing Captives

In an 1884 lecture at the Smithsonian Institution, physician Charles K. Mills recounted some of the hazards of scientific investigation. Delivered as part of a prestigious series dedicated to "the promotion of medical science," the lecture enumerated the range of diseases (albuminuria, temporary nervous collapse, insanity) that might afflict the overworked scientist. Valiant scientific labors—"assiduous work with the microscope, steady concentration upon mathematical and engineering problems, and the laborious collection and comparison of data"—threatened the fragile investigator with a host of psychopathologies. "To some minds," Mills summarized, "scientific work has a fascination which becomes a source of peril; the worker becomes a willing slave to tasks which are often of his own making."[1] Here the scientist is at once master and servant, a "willing slave" who conditions his own captivity.

Appearing as both agent and victim, the captive scientist exemplifies an understanding of civilized manliness that had broad support in late nineteenth-century America. An 1885 editorial in the *Nation* echoed Mills's concerns. The "decrease of country population, the increasing haste and excitement of city life, the increasing demand of extensive [education] at the cost of intensive education, [and] the universal demand for luxury and pleasure" all threatened American men. Particularly at risk, author G. Stanley Hall proposed, were "brain workers" who perform, "according to the recent estimate of a physiologist, about eight times as much mental work as a manual laborer."[2] From Theodore Roosevelt's endorsement of the strenuous life to George Beard's treatises on neurasthenia, men of cultivation were regularly depicted as both especially powerful and especially endangered, uncommonly susceptible to the effects of overcivilization. While some white women managed to engage norms and ideals of voluntary captivity (almost exclusively as nobly suffering mothers or missionaries), most accounts concentrated on the special vulnerabilities of educated, professional men.[3] To understand the ethic of self-sacrifice evident among late nineteenth-century scientists, we must first appreciate the era's broader preoccupation with voluntary suffering.

To elucidate this context, this chapter pulls back from a focus on the peculiar vulnerabilities of the sacrificial scientist to chart the general condition of the voluntarily submissive self in nineteenth-century America. Weaving together religious, economic, legal, and intellectual history, it tracks the figure of the willing captive from the antebellum period through its appearance in Mills's 1884 remarks. A quivering ganglion of rights, faculties, and properties, this willing captive embodied the larger tensions of the volatile century. At the heart of these tensions was a perspective on voluntarism with roots in seventeenth-century liberal political theory. As we trace these roots into the late nineteenth century, two salient attributes of the voluntary, liberal self become clear: first, this self was founded on a certain obligation—namely, the imperative to enter into exchange with larger social bodies; and second, this self was sexually, racially, and nationally distinct, defined in opposition to bodies characterized by dispossession and exclusion from civic participation.[4] Both features of the voluntary self would prove consequential to the operation of self-sacrifice in science.

Because this chapter departs from the book's general attention to science and scientists to move broadly across histories of religion, political philosophy, and law, a brief synopsis of its central assertions may prove useful. The chapter's organization proceeds chronologically from the religious revivals and political revolutions of the late eighteenth century through the widespread economic, legal, and social upheavals occurring during and after Reconstruction. In the first section, I explain that as relations between will, selfhood, embodiment, and property fluctuated in the decades immediately preceding the Civil War, ownership of the self came to be requisite to participation in civil society. As a result, in courtrooms and churches, at strikes and séances, antebellum Americans disputed which beings (corporations, children, Indians, Negroes, prisoners, wives, the insane) might be said to be fully in possession of their faculties and capacities and thus which beings might obtain the full set of social privileges accorded to the true person. Voluntary suffering, which at once presumed and reinforced the condition of self-ownership, played an especially important role in fixing the meanings of volition and compulsion and thus in reconfiguring debates over selfhood, personhood, embodiment, and civic participation.[5]

These debates, I show, took on renewed importance during and after the war between the states, as Americans clashed anew over relationships between individual selves and the social and political bodies of which they were a part. Although Emancipation reformed the most obvious and pressing inequities of property and self-determination, it did not resolve the inherited conundrum of just how much suffering a reasonable self must—or ought—to endure. Indeed, dissent over suffering shifted from chattel slavery to other forms of compulsion and coercion: the directives of an all-powerful God; the imperatives of natural selection; the dictatorship of feeling, habit, and sentiment; and the inexorable advance of industrial capitalism. Conflict over the unwilling subjection of southern slaves, resolved by Union victory, was displaced by conflict over the value and necessity of *willing* subjection: of voluntary subjugation to the demands of God, nature, nation, and industry. As Mills's popular 1884 lecture "Mental Over-Work and Premature Disease among Public and Professional Men" demonstrates, in the years following Reconstruction, the line between reasonable, willing submission and unreasonable, pathological forms of captivity turned out to be anything but clear.

IN SOME SENSES, Mills's assessment of diseased relationships to science simply layered a medical lexicon of health and illness over old theological disputes. How individuals might best achieve communion with the immortal body of God had long been subject to intense scrutiny in the diverse traditions of American Protestantism. For these children of Reformation, the question centered on the extent to which deliberate effort might affect personal salvation: whether one's labors would influence deliverance or whether the doctrine of predestination rendered such labor irrelevant. The theological conundrum of free will occupied the New World faithful no less than their European predecessors.

Suffering focused the dilemma of predestination in a distinct manner. For the innumerable American Protestants who studied John Foxe's *Book of Martyrs* alongside the family Bible, stories of their fellows roasted over spits, devoured by wild beasts, or thrown into vats of boiling pitch posed a disconcerting problem. Was suffering a sign of punishment for sin or an opportunity to improve Christian standing? If suffering might actually help gain access to heaven, just how much should it be embraced? Ought one seek out boiling pitch or ravenous beasts? If exertion increased the likelihood of salvation, might particularly painful, spectacular effort—bloody martyrdom—guarantee God's favor? Was sacrifice, as Cotton Mather suggested, a "Holy Skill" that might be honed through intentioned endeavor?[6]

While stopping short of suggesting that God could be coerced by an individual's earthly activity, a number of colonial theologians did emphasize the role of intentional preparation in attaining redemption. After all, such a perspective was in keeping with their expectations of the new continent: wholesale resignation to God's will would run counter to most colonists' assurance of their special covenant with the divine. As historian Perry Miller explains, the theological reform that took hold in Puritan New England renewed the covenant as an agreement based on a kind of mutual obligation. For his part, God agrees to provide both salvation to believers and the capacity for belief itself. As for man, he need only affirm that he will accept the grace that God has given. "If he can believe, he has fulfilled the compact; God then must redeem him and glorify him."[7] In effect, God willingly binds himself to a contractual exchange. He in no way compromises his dominion in doing so, for he is not "compelled by any force of reason or necessity." Through the doctrine of covenant expressed by colonial theologians, God becomes "a God chained—by His own consent, it is true, but nevertheless a God restricted and circumscribed—a God who can be counted upon, a God who can be lived with."[8]

Colonial clergy thus crafted a moderate position on the spectrum between autonomy and predestination, one that preserved some room for self-determination. Moral freedom rested with the individual, but an individual subjected to the dictates of a higher power. Isaac Backus, one of the most influential Baptist ministers of the late eighteenth century, summarized this perspective simply when noting that genuine freedom meant the "freedom to observe the divine will."[9] Working from this middle ground, even a well-known proponent of predestination like Jonathan Edwards stressed man's obligation to work deliberately toward acquiring God's grace. As Edwards instructed his congregation in the winter of 1734, "If we would be saved, we must seek salvation. For although men do not obtain heaven of themselves, yet they do not go thither accidentally, or without any intention or endeavors of their own. God, in his word, hath directed men to seek their salvation as they would hope to obtain it."[10]

In insisting on the necessity of intentional effort, Edwards presaged the perspective brought forward by revival movements of the early nineteenth century. In manifold ways, these movements continued the project of accommodating the doctrine of predestination to the American context. As a result, by the opening decades of the nineteenth century, the strict determinism of orthodox Calvinism had been superceded by a focus on the liberty of each believer. While liberty continued to suggest (as it had for eighteenth-century clergy) the freedom to submit to a higher power, the significance of individual will in relations to God moved increasingly to the fore. Northern Protestants in particular emphasized the individual's ability to achieve his own salvation, deliberately distancing themselves from the servility of the Catholic immigrant and the captivity of the southern slave. One could *choose* to be saved, these northerners insisted, and labor accordingly.[11]

The burgeoning emphasis on free will and individual liberty evident in antebellum revivalism was strengthened by a concurrent reorganization of the structures of democratic participation. At the time of the nation's founding, the right to participate in the institutions of governance resided in ownership of productive property; accordingly, the condition of individual political liberty was embedded in possessed wealth. John Adams expressed the bond between property ownership and liberty in the spring of 1776, when he suggested that men "wholly destitute of property" are "too dependent upon other men to have a will of their own."[12] Yet as Eric Foner has shown, revolutionary movement gradually transformed the presumed connection between property ownership and participation in the civic body and, with it, the character of the free self. No longer was political freedom automatically associated with the possession of productive property, with the holding of land or currency. By the eve of the Civil War, most states granted suffrage not only to the propertied elite but to all adult white men.[13]

It is crucial to note, however, that although this revolutionary shift sundered the link between land and liberty, it did not actually sever the connection between *property* and liberty. Property continued to be fundamental to freedom; but now possession of the *self,* rather than possession of dispensable capital, became the politically consequential form of proprietorship. In the years leading up to the war, self-ownership came to be the defining condition of civic participation. This transformation set the stage for a reorganization of selfhood and, concurrently, of the self's membership in larger associations, be they divine, national, or—later—scientific.

The conceptual foundations of this revised vision of selfhood and civic participation had been laid out more than a century earlier by property's most influential philosopher, John Locke. In the *Second Treatise* of 1690, Locke explained that "every Man has a *Property* in his own *Person.* This no Body has any Right to but himself."[14] Man's proprietal relation to himself, Locke proposed, forms the basis of consensual political union, the ground—indeed, the only ground—from which men might enter agreement with other agents. These agents need not be other sentient beings; individuals might just as easily contract with nature or with the state.[15] In each case, however, the liberal subject, the self with property in his own person, chooses how and when to barter his capacities.

Like the colonial Christian, even the willful Lockean self remains, in the final analysis, a subject—the servile recipient of God's creative gift. Locke proposed a limited sense of possession in this regard, for ultimately life was given by another, absolute owner: "a Man, not having the Power of his own Life, cannot, by Compact, or his own Consent, enslave himself to any one, nor put himself under the Absolute, Arbitrary Power of another, to take away his Life, when he pleases."[16] The Lockean concept of self-ownership, the centerpiece of liberalism, is thus more akin to a loan or a trust than outright, autonomous ownership. Even the most independent individual possesses himself only provisionally. Locke's negotiation of this point indicates another accommodation of the conundrum inherent in created free will.

The Lockean vision of property-in-the-person, which presupposes some distinction between the deliberative self and "his" body, was not the only conception of the person available in the antebellum period. Discussions of privacy, for example, presumed no similar split between active self and passive property: the person was inescapably embodied and the body inescapably personified. Other commentators treated the person as defined by bodily necessity rather than as something existing before it; still others advanced the sacred, inviolable holism of body and spirit.[17] The rise of the proprietal self—and of the political and economic institutions which gave it material form—hardly eradicated competing understandings of the individual. The "I," Hegel remarked in 1817, remained something of a philosophical "void," a "receptacle for anything and everything."[18] Yet the proprietal self became the dominant frame of antebellum personhood. American intellectuals, Miller holds, "were not so much the beneficiaries of Locke as they were his prisoners."[19] As religious liberalization increasingly emphasized the role of individual volition in the attainment of salvation, so, too, the shifting structures of democratic governance enshrined the self-possessed man as the relevant unit of the polity.

This proprietal self came to uphold the ethos of self-sacrifice claimed for science by commentators like physician Charles K. Mills and yellow fever experimenter Walter Reed. Their varied remarks make clear that voluntary suffering for science rests on a prior, Lockean principle of property-in-the-person: only that which is owned may be forfeited. In valorizing self-sacrifice, late nineteenth-century scientists would both presuppose and enliven the proprietal constitution of the self. As subsequent chapters explain, just as self-ownership came to define membership in the polity, so, too, this liberal concept of selfhood would come to condition membership in the imagined body of science.

Given its eventual importance to scientists' self-sacrifice, two aspects of the antebellum proprietal self merit emphasis here. First, it is worth noting that the volitional self is founded on a certain obligation: paradoxically, self-ownership compels exchange. Locke and other architects of liberalism famously argued that each "free" individual engages in social relations as a kind of barter. The individual voluntarily surrenders his original liberties in exchange for the securities provided by the group. This exchange is the social contract envisioned by Locke, Hobbes, and Rousseau. "Individual self-protection is the problem that has to be solved in the state of nature," summarizes political theorist Carole Pateman, "and the solution is the contract."[20] All social relations may be seen as an "endless series of discrete contracts."[21]

The crucial point for present purposes is the necessity of this contractual activity: as political philosopher C. B. Macpherson demonstrates, one consequence of Lockean property-in-the-person is that social interaction cannot be conceived as anything other than a form of exchange. "Since the individual is human only in so far as free, and free only in so far as a proprietor of himself, human society can only be a series of relations between sole proprietors, i.e. a series of market relations."[22] To posit individuals as proprietors of their own capacities can only imply that they are free to sell or trade those capacities for subsistence. The self designated as owner of his person and his attributes is obliged to participate in a ceaseless procession of exchanges. This compulsion to exchange gives rise to the defining paradox of the proprietal self: the individual's

possession of his bodily capacities (including the capacity for labor, love, reproduction, and so on) is premised on his relinquishing of those same capacities,
on judiciously entering them into social circulation. As we will see momentar-
ily, while the voluntary self is always and everywhere shackled to the burdens of
the exchange, the form and strength of these shackles shifted markedly in the
transition from slavery to free labor—just as self-sacrifice came to the fore in
discussions of science.

The obligations attending proprietal selfhood are just one of its noteworthy
attributes. Of equal relevance to later versions of self-sacrifice is that the proprietal self is always defined relationally—in accordance with what it is not. The self
depended on comparison with others: possession was delimited by dispossess-
sion, liberty by captivity. As recent critics of social contract theory have argued,
the proprietal self developed by canonical political philosophers acquired its
distinctive autonomy and independence only in contrast to the subordination
and dependence of others. In the words of philosopher Charles W. Mills, the social contract has not been a contract "between everybody . . . but between just
the people who count, the people who are really people."[23] Antebellum writers
themselves noted the constitutive, foundational character of these relationships. The Richmond Enquirer put the point succinctly in 1856: "Freedom is not
possible without slavery."[24] In innumerable legislative and judicial battles, the
boundary of the voluntary self was drawn and redrawn according to the changing circumstances of the unfree. For instance, assumptions about manly self-
determination are manifest in antebellum disputes over wifely subservience;
similarly, legal contests over insanity shaped standards of "sound mind."[25] In the
years leading up to the Civil War, which beings were to be counted as fully self-
possessed persons was a matter of persistent and bloody conflict.

Conflict over the relational character of personhood is evident in an 1844
U.S. Supreme Court case, in which the Court considered whether a being might
be treated as legally culpable even if it possessed no individual will. The case
began when a ship captured from pirates by a U.S. Navy schooner in 1840, the
brig Malek Adhel, was condemned and sold under existing antipiracy laws. The
original owners of the ship objected to the sale, and the case went to litigation.
In its ruling on the ship's impoundment, the Court affirmed the owners' inno-
cence in the case yet also upheld, over the protestations of the owners, the legality of the seizure. The Court reasoned that although the ship had willed no
malfeasance (since the wooden brig obviously lacked intent), it might still be
held accountable for injuries. The Court cited an 1819 ruling by Chief Justice
Marshall to this effect: "This is . . . a proceeding against the vessel for an offence
committed by the vessel; which is not the less an offence . . . because it was committed without the authority and against the will of the owner." The Malek Adhel
was thus not an entirely passive "vessel," but nor was it entirely self-possessed

and autonomous. Cases such as *U.S. v. the Brig Malek Adhel* reveal that the attribution of agency, responsibility, and self-determination was always achieved in flexible relationship with passivity, dependence, and submission.[26]

Of course, nowhere in antebellum America were the constitutive exclusions of proprietal selfhood more apparent than in the institution of slavery. The central distinction between freeman and captive in the slaveholding era was one of property-in-the-person: the free self maintained willful ownership of his body. In contrast, the captive was forcibly stripped not only of ownership of the body but also, in the eyes of the state and slaveholding whites, of volition itself. The first theft inflicted on African peoples, Hortense J. Spillers writes, was the "severing of the captive body from its motive will."[27] Barred from the foundational right of self-ownership, the slave, marked as "black" regardless of skin color, was said to be *only* flesh: the "antithetical embodiment of pure will."[28] Dispossessed of bodily ownership (the origin of the subject, in Locke's account), the captive body was thus dispossessed of the selfhood requisite for civic participation. Years later, the residue of this violent history of dispossession—of distinctions between those allowed to *possess* bodies and those allowed only to *be* bodies—would continue to inform the realization of self-sacrifice.[29]

The relational distinction between the free, proprietal self and the captive, dispossessed body acquired physical incarnation in the act of suffering. In the slaveholding era, bodies said to be impervious to pain were said to be lacking the necessary faculties of the free person; slaves, in turn, were said to be impervious to pain. The disavowal of sensitivity to pain was tantamount to the denial of humanity itself. By forcing slaves to sing, dance, and play cards in markets and "strike up lively" in coffles, slavers sought to perform the absence of selfhood requisite to enslavement in a liberal democracy.[30] Pro-slavery medical experts contributed to this project, generating study after study on the inhuman hardiness of the slave. As Dr. A. P. Merrill wrote in 1855, "They submit to and bear the infliction of the rod with a surprizing degree of resignation, and even cheerfulness. . . . They differ from their white masters in no one particular more than this."[31] While opponents of slavery diverged from apologists like Merrill in their conclusions, many reinforced slavers' assumptions of differential racial susceptibilities to pain. Even some of the staunchest white abolitionists reproduced claims of black insensibility, seeing it not as justification for enslavement but as evidence of God's compassion. Lydia Maria Child, for example, applauded the "merciful arrangement of divine Providence, by which the acuteness of sensibility is lessened when it becomes merely a source of suffering."[32] Whether specially blessed or cursed by the Creator, enslaved men and women were presumed to lack the characteristic vulnerabilities and sensitivities of the fully self-possessed.[33]

It is important to note, however, that the boundary demarcating the willful self from the will-less captive was never simple and stable; there was no absolute bifurcation between sensitivity, freedom, and autonomy, on the one hand, and complete inhumanity on the other. Rather, the contradictory demands of race-based slavery generated ambiguous, fractional definitions of selfhood at every turn. The slave owner's need to simultaneously affirm and deny the slave's humanity relegated some bodies to what Spillers terms a "cultural vestibularity"—near, but not altogether inside, the structures of governance engendered by full personhood.[34]

This vestibularity was made starkly apparent at the Constitutional Convention of 1787. At the convention, the task of setting apportionment for purposes of representation and taxation highlighted the uncertain, relational nature of proprietal selfhood in an immediate and inescapable manner. Delegate Gouverneur Morris spelled out the contradiction: "Upon what principle is it that slaves shall be computed in the representation? Are they men? Then make them Citizens & let them vote? Are they property? Why then is no other property included?"[35] The relativity of freedom and self-ownership is indicated in the delegates' infamous compromise: to count "the whole number of free persons, including those bound to service for a term of years, and excluding Indians not taxed, three fifths of all other Persons." With this decision, constitutional framers entered slaves in the vestibule of the nation: as partial, "three fifths," persons.

The *partial* personhood of the racialized body aptly evokes both the fractional and biased senses of that word: the same unstable relations between will, selfhood, embodiment, and property that shaped entry into antebellum civic life also influenced beliefs about credibility. Consider, for example, an 1854 case in which the Supreme Court of the state of California heard an appeal considering whether the appellant, "a free white citizen," could be convicted of murder on the testimony of nonwhite witnesses. Existing law stated that "No Black or Mulatto person, or Indian, shall be allowed to give evidence in favor of, or against a white man." The court was thus called to determine whether witnesses of "the Mongolian type" were to be considered similarly. Overturning the lower court's judgment, the higher court ruled to exclude the testimony of "Chinese witnesses." Asserting that "the Mongolian" was part of "a race of people whom nature has marked as inferior, and who are incapable of progress or intellectual development beyond a certain point," the court found Chinese witnesses inadequate to the demands of testimonial credibility. (In reversing the lower court's decision on the same question, Judge Charles J. Murray noted that the earlier legislation dated to a time when ethnology was not yet at "that high point of perfection which it has attained by the scientific inquiries and discoveries of the master minds of the last half century.")[36]

Again, the ruling did not entirely remove Chinese witnesses from the realm of the social: the Mongolian remains a person, as does the Black, Mulatto, and Indian. Yet the ruling reinforces the cultural vestibularity Spillers describes: one might well be a person but still not a credible self. Deemed naturally inferior by the unembodied "minds" of ethnological science, "the Chinese," according to the court, lacked the capacities for testimonial credibility requisite to full civic participation. Bodies might be partial persons—possessed of some, but not all, of the attributes accorded to the archetypal free liberal subject.

It bears repeating that the Chinese in this case were rendered partial persons only in and through relation to those installed as impartial—as full, free participants in civic affairs. Testimonial impartiality, like self-possession, was established relationally. As the 1854 ruling spells out explicitly, testimonial credibility and the rights and privileges it provided rested on comparative standards of racial difference. By the use of this term "black," the Court wrote, "we understand it to mean the opposite of 'white,' and that it should be taken as contradistinguished from all white persons":

> In using the words "no black or mulatto person, or Indian shall be allowed to give evidence for or against a white person," the Legislature, if any intention can be ascribed to it, adopted the most comprehensive terms to embrace every known class or shade of color, as the apparent design was to protect the white person from the influence of all testimony other than that of persons of the same caste. The use of these terms must, by every sound rule of construction, exclude every one who is not of white blood.

Based on this logic, the highest court in California prohibited the participation of Chinese witnesses in the central institution of democratic governance. An association between whiteness and testimonial credibility was thus not merely an effect of racial disenfranchisement but part of the very means by which the law perpetuated its disenfranchising work.[37] Similarly, when the constitutional framers tallied slaves as three-fifths of free persons, they also, just as surely, called forth the white citizen as a *whole* self, one granted unimpeded access to the body politic. For the self with such full possession of his person and its attributes, the body could be rendered juridically irrelevant and hence effectively invisible. And as these examples attest, the invisibility of some bodies came at the cost of others.[38] As subsequent chapters will show, the traditions of partial personhood that limited participation in the public body of the nation would also shape membership in the emerging body of science.

THE UPHEAVAL commenced by the siege of Fort Sumter further shook the ground of suffering, self-possession, and civic participation tilled by earlier religious and legal trends. As the war deepened, white commentators on both sides of the Mason-Dixon line struggled to apprehend the piles of gangrenous limbs and blackened corpses piling up around them. Such colossal losses cried out for explanation, and northerners and southerners alike sought this purpose in the inscrutable but sure wisdom of divine reason. Claiming the carnage of war to be part of God's larger plan, both northern and southern nationalists attempted to confer collective immortality on individual death.

Conferring this collective immortality—to make an action appear as "sacrificial" rather than merely "suicidal"—required a reorganization of time: the postponement of immediate and observable good sense for the sake of some envisioned future. What might first seem to be an unreasonable forfeiture of life and limb became a reasonable offering when premised on an imagined communion between past, present, and future citizens. To ensure this communion, some time must elapse between the demise of the individual soldier and the unification of the nation; otherwise, the soldier's death would fail to acquire its purported significance: the preservation of the nation's longevity. Yet this span of time must not be infinite, or the individual's death would seem merely purposeless, mad.[39] This experience of time—present submerged in a distant but not *too* distant future—secured the deferrals of immediate and obvious interest necessary for the continuation of war.

As in spiritual salvation, this was a deferral in both senses of the word: a delay and a yielding. The suspension of apparent understandings of reasonable action (does it really make sense to run headlong into a line of cannons?) was enabled by subjugation to the obscure but assured will of God. Faith, acquired in homes and churches long before the first shots rang out at Sumter, sustained conviction in suffering's larger purpose, transforming senseless slaughter into a sensible, if delayed, exchange. Once metaphysical conviction was transferred to the mystical body of the nation, a willingness to suffer and die in its name became not only permissible (rather than blasphemous) but expected. Put another way, habits of religious consecration were not secularized so much as the ostensibly secular realm of the nation was sacralized.[40]

Such sacred nationalism appears to have been particularly consequential for white women, who played a central role in its promotion. While most combatant men drew continued commitment to the war effort from the men fighting beside them, noncombatant women tended to call on the assurances of divine providence, particularly as the war dragged on. Portrayed as endowed with a "positive love of self-sacrifice," women enflamed one another to further suffering. And just as earlier Protestant theologians disputed the relations between voluntarism and the attainment of salvation, so, too, wartime women writers waffled on exactly what role individual effort played in the realization of national aims. If the ends of sacred nationhood were already ordained, what role might each individual play in its attainment? Could one actively speed the completion of the divine plan by offering up yet another brother or son?[41]

Differing perspectives on the role of human will in suffering are evident in two texts from the spring and summer of 1863. As civilian deaths from starvation spread through the south, one Richmond periodical spoke of the comfort to be found by surrendering to God's benevolent directives:

But e'en if you drop down unheeded,
What matter? God's ways are best:
You have poured out your life where 'twas needed,
And He will take care of the rest.[42]

In lieu of such pious abdication, another periodical published in 1863 advocated more eager, active effort—an exaggeration rather than submergence of the willful self. "A Call to My Country-Women" urged northern readers to seize their status among the suffering elect energetically:

Take not acquiescently, but joyfully, the spoiling of your goods. Not only look poverty in the face with high disdain, but embrace it with gladness and welcome. The loss is but for a moment; the gain is for all time. Go farther than this. Consecrate to a holy cause not only the incidentals of life, but life itself. Father, husband, child,—I do not say, Give them up to toil, exposure, suffering, death, without a murmur;—that implies reluctance. I rather say, Urge them to the offering; fill them with sacred fury; fire them with irresistible desire; strengthen them to heroic will. . . . Count it all joy that you are reckoned worthy to suffer in a grand and righteous cause.[43]

Here the will is toughened and enlarged rather than surrendered. The true Christian stimulates the desire for sacrifice, pushing her loved ones toward further "toil, exposure, suffering, and death." While both texts stress the holy character of wartime suffering, they differ in the relative weight given to individual volition in fulfilling God's plan.

The question of will was of particular consequence in the north, where abolitionism had long presupposed revulsion for pain and positioned enthusiastic endorsements of suffering rather precariously. Given their long-standing criticism of violence, many New England abolitionists embraced the brutality of war with astonishing rapidity and eagerness. "I can hardly help wishing that the war might go on and on," wrote Charles Eliot Norton to a friend in the wake of casualty reports from Shiloh, "till it brought us suffering and sorrow enough to quicken our consciences and cleanse our hearts."[44] Norton's simultaneous opposition to the violence of slavery and support for the violence of war points to the antislavery movement's elevation of voluntary suffering and the concept of proprietal selfhood that such volition presupposed. As historian George Frederickson explains, "While seeking to relieve the suffering of the slave, the abolitionist . . . welcomed it for himself."[45] For northern liberals such as Norton, the key to distinguishing beneficial, cleansing violence from undesirable, immoral brutality was to be found in the voluntarism of the sufferer. In the life of the slave who lacked freedom of deliberation, suffering was despicable and debasing; in the life of the self-possessed, white Christian, suffering was noble and ennobling.

The importance of volition in sorting types of suffering and types of suffer-ers is evident in Ralph Waldo Emerson's popular poem "Voluntaries." First pub-lished in the *Atlantic Monthly* in October 1863, the verse honors the charge of a South Carolina battery by the young Bostonian Robert Gould Shaw and members of the first "colored" regiment. The regimental soldiers themselves, described by Emerson as descendents of "Afric's torrid plains," are said to have "avenues to God/Hid from men of Northern brain." Their avenues to God, however, are indirect—routed through the will of the youthful white colonel. These "Foundling[s] of the desert" might partake of military heroism, but only as steered by the "gen-erous chief" who leads those "willing to be led." The youthful chief, for his part, has no such intermediary. Duty speaks to Shaw directly:

> In an age of fops and toys,
> Wanting wisdom, void of right,
> Who shall nerve heroic boys
> To hazard all in Freedom's fight,—
> Break sharply off their jolly games,
> Forsake their comrades gay
> And quit proud homes and youthful dames
> For famine, toil and fray?
> Yet on the nimble air benign
> Speed nimbler messages,
> That waft the breath of grace divine
> To hearts in sloth and ease.
> So nigh is grandeur to our dust,
> So near is God to man,
> When Duty whispers low, *Thou must,*
> The youth replies, *I can.*[46]

These lines reproduce the two characteristics of proprietal selfhood discussed thus far, characteristics that reappear in the ethos of self-sacrifice in science. First, note that volition, which defines the patriotic volunteer, is attributed un-evenly and relationally. The regimental soldiers display some capacity for hero-ism, but only when encouraged by the generous chief. Occupying the vestibular personhood described by Spillers, these soldiers have only partial access to ideals of voluntary suffering. In contrast, Shaw, born of "proud homes" and raised amidst circles of "sloth and ease," appears to hold willful autonomy in greater measure. His is a direct relationship to the divine, an unmediated ex-change. Second, note that to the extent that Shaw possesses independent voli-tion, it is confirmed by a willingness to endure "famine, toil and fray." Just as his freedom and self-possession enable his ability to "volunteer" himself, so his choice to suffer verifies the nature of his self-ownership, since only the pos-sessed may be voluntarily hazarded. Of course, in keeping with the conventions

of Lockean self-possession, even the volunteer's will falls subject to God's external directives; ultimately, the willful self—the determined I of the *I can*—is on
loan from a higher power. Nevertheless, quite unlike the vestibular members of
the colored regiment, the affluent white volunteer has a whole self to lose.

THE SUFFERING SELF that flowered amid the war's devastation did not wither
with the end of armed conflict. Rather, its defining paradoxes and exclusions
were sustained and intensified by later developments. Among the most important of these changes were the sweeping industrial changes of the post-
Reconstruction era, which at once eroded the material bases of self-determination
and installed the wage-earning individual, rather than the family or the community, as the locus of economic responsibility. As farms depopulated, cities
swelled, and households ceased to be self-sustaining units of production, more
and more Americans were bound to the machinery of industrial capitalism.
Growing numbers of itinerant laborers were drawn into mines, mills, and facto-
ries; by 1900, most white men were working for wages or salaries. Declining self-
employment coincided with the rise of more overtly coercive and obligatory
arrangements of work, including debt peonage, forced pauper labor, compulsory education, and a spreading convict lease system. Even members of the
comfortable classes grappled with increasingly sedentary and bureaucratized
lives. Between 1880 and 1920, the number of salespeople, clerks, government
employees, professionals, and technicians in the country ballooned from 756,000
to 5.6 million.[47]

Taken together, these shifts rendered obsolete understandings of individual freedom based on control over daily work. Increasingly, the terms of
workers' everyday lives—sexual behavior, religious observation, language use,
alcohol consumption, and so on—were subject to intensified managerial regulation. As wages and contracts supplanted whips and coffles, the relations between liberty and coercion were newly in question. Sociologist William Graham
Sumner declared in 1883 that the "modern free system of industry" offered
"chances of happiness indescribably in excess of what former generations have
possessed" yet also stripped men of whatever guarantee they might once have
possessed "that they should in no case suffer."[48] The lash was effectively inter-
nalized: the "free(d) individual," Saidiya Hartman concludes of the period, "was
nothing if not burdened, responsible, and obligated."[49]

To say that late nineteenth-century selves were subject to renewed forms of
compulsion and obligation is not to diminish the emancipatory events wrought
by centuries of violent and nonviolent resistance. By 1870, the ratification of the
Thirteenth, Fourteenth, and Fifteenth amendments had abolished the legal cor-
nerstones of racial enslavement: the Thirteenth by prohibiting slavery and involuntary servitude (except for those "duly convicted" of crime); the Fourteenth
by rendering all "persons" as citizens, with equal rights to life, liberty, property,

due process, and protection; and the Fifteenth by securing that rights would not be abridged on the basis of race, color, or prior enslavement. But while these amendments revoked some of the Constitution's most egregious legal contradictions on the matter of liberty (such as the three-fifths rule of representational apportionment), they did not address the fundamentally indebted character of the free, proprietal self in a capitalist society. In fact, the economic, political, and intellectual metamorphoses of the last quarter of the nineteenth century merely intensified the ambiguities of volition and coercion contained in the liberal subject. This intensification can be made more apparent by considering examples of three different post-Reconstruction examples of vestibular personhood: wives, Chinese immigrants, and corporations.

Tensions between liberal ideals of contractual freedom and the bondage of married women had long been apparent to reformers, and a series of legal improvements during the mid-nineteenth century gave women who worked for wages increasing economic independence. These reforms troubled the prior supposition that free husbands and their wives formed one political unit, with wives merely an extension of individual men. Married women's inability to own and transfer productive property without entering into complex trusteeships had been one of the dominant concerns of the Declaration of Sentiments signed by attendees of the 1848 Seneca Falls Convention. Continuing agitation over the issue eventually resulted in the passage of married women's property acts in several states. Earnings laws, many of which explicitly stated women's ability to enter into independent contracts, followed soon after; and by 1887, two-thirds of the states had instituted such laws.[50]

Yet these piecemeal and labyrinthine reforms in marriage law did not resolve the question of woman's self-ownership as the Thirteenth Amendment had for freedmen. On the contrary, several prominent congressmen offered their support for Reconstruction reform on the principle that emancipated black men would then join white men in assuming their rightful command over wives and children. According to one Republican from Iowa, the right to "personal liberty," the right of "a husband to his wife," and the right of a "father to his child" were the "three great fundamental natural rights of human society."[51] As advocates for women's political enfranchisement readily recognized, the right to hold and transfer productive property did not necessarily translate into the full *self*-possession requisite for unfettered civic participation.

Indeed, as the nation assumed new priority in establishing the basic rights and privileges of citizens, the continued vestibularity of women only became more pronounced. Prior to ratification of the Fourteenth Amendment, states held the power to determine women's political participation: the Constitution set out no voting requirements and thus placed no restrictions on the extension of suffrage to women. With the ratification of the Fourteenth Amendment, however, federal law explicitly and for the first time limited representation to *males* over twenty-one years of age (it also continued to exclude representational apportionment for "Indians not taxed").[52] Postwar legal reform was thus decisive not only in reinvigorating self-ownership as the fundamental principle of civic participation but also in restricting access to such participation to men. Male control of women's sexual and economic capacities was maintained within the bonds of marriage, denying women the fullest sense of property-in-the-person.[53] In other words, even as rights to self-possession and civic participation were extended beyond previous racial exclusions, women's claims to self-possession remained one of persistent partiality in the wake of Reconstruction. As will be seen, women's vestibular relationship to the nation would be reproduced in their vestibular relationship to the emerging community of scientists.

Postwar constitutional reform affected relations among will, embodiment, and civic inclusion in other realms as well, perhaps nowhere more visibly than in debate over the restriction of Chinese immigration. After 1865, the Chinese immigrant replaced the southern slave as the icon of bondage, the quintessentially coerced body against which to define the liberties enjoyed by others. Accordingly, delineating the character of the Chinese immigrant's voluntarism became a common endeavor.[54] Immigrant Chinese laborers, asserted one former minister to China in 1876, "are more absolute slaves than ever the negroes of the South were."[55] Ulysses S. Grant similarly stressed the coercion of debt-bound immigrants in his 1874 presidential address, devoting particular attention to women forced into prostitution: "I call the attention of Congress to a generally conceded fact that the great proportion of Chinese immigrants who come to our shores do not come voluntarily . . . but come under contracts with headmen who own them almost absolutely. In a worse form does this apply to Chinese women. Hardly a perceptible percentage of them perform any honorable labor, but they are brought for shameful purposes."[56] As Grant's address suggests, compulsory prostitution, a particularly vivid specter of freedom lost, haunted debates over Chinese immigration throughout the 1870s and 1880s, with commentators arguing over just how voluntary migration might be.[57]

Indeed, a crucial feature of postwar debates over Chinese immigration was the contested definition of individual consent. The enforcement of the 1875 Page Law, for example (the federal government's first explicit exclusion of specific immigrant groups), turned on the government's ability to validate the free will of individual immigrants. The American consul in Hong Kong, charged with the implementation of the law from 1875 to 1877, was called to adjudicate the veracity of women's answers to questions such as "Have you entered into contract or agreement with any person or persons whomsoever, for a term of service within the United States for lewd and immoral purposes?" and "Do you wish of your own free and voluntary will to go to the United States?"[58] As the nation superceded the state as the adjudicator of enfranchisement with the ratification of the Thirteenth, Fourteenth, and Fifteenth amendments, the character of volition was now subject to federal determination. Who might be said to possess a "free and voluntary" self became a matter of national concern.

In the years after Reconstruction, even nonhuman bodies were caught up in this concern, as demonstrated by the opinion of the U.S. Supreme Court in the influential *Santa Clara County v. Southern Pacific Railroad Company* (1886). At first glance, the decision appears far removed from matters of selfhood discussed thus far: the dispute centered on how fences along railway lines were to be valued and assessed. The defendants argued against some forms of taxation on the ground that states were forbidden to deny to any person within its jurisdiction the equal protection of the laws. Corporations, the defendants argued, were persons, too. Chief Justice Waite refused to hear argument on that point, noting simply that in the opinion of the Court corporations did fall within the jurisdiction of the amendment. With this decision, the Court installed the corporation as a body, a corpus, possessed by its shareholders. As a person without self-propriety, the corporation is ostensibly a person in a condition of slavery; and the shareholders, in turn, appear to be in violation of the Thirteenth Amendment to the Constitution. The Court did not say whether the corporation's status as a *willing* subject exempted it from the amendment's prohibition on "involuntary servitude."[59] The case makes clear that questions of coercion, will, and self-possession—can a person be at once free and voluntarily servile?—remained undecided well after Emancipation. Postwar constitutional reform created new bottles for old theological, legal, and political ferment.

THE REORGANIZATIONS of selfhood, proprietorship, and collective participation instantiated by postwar economic and legal changes drew sustenance from attendant intellectual and social developments. Chief among these was the dissemination of principles of Darwinism. Initial American reception of the 1859 *Origin of Species* was muted by preoccupations with mounting sectional conflict. Yet even in the first years of the war, evolutionary theory was used to justify the individual's submission to the larger collective. For example, an 1861 article by Charles Eliot Norton, written in the wake of massive casualty reports from Bull Run, argues that history, faith, and nature equally dismiss the "little value of the individual in comparison to the principles upon which the progress and happiness of the race depend." Given the relatively low worth of the individual, Norton argued, the opportunity to suffer on behalf of a transcendent, immortal entity is "a gift." In Darwinism, Norton found fresh justification for the larger deferral of time on which all imagined communions depended. In nature as for the nation, the mortal individual's present existence is exchanged for the immortal, collective future: "Nature is careless of the single life. Her processes seem wasteful, but out of seeming waste, she produces her great and durable results. Everywhere in her works are the signs of life cut short for the sake of some effect more permanent than itself." Voluntary subjection—to "die for truth, to die open-eyed and resolutely for the 'good old cause'"—becomes inescapable, the natural telos of Christian nationalism.[60]

As Norton's remarks show, the step from virtue to necessity was so narrow as to be nonexistent: moral reasoning seemed to spring from natural law. Panegyrics to the valor and nobility of suffering merely exaggerated the assumption, pandemic in the wake of Darwinism, that all progress required some degree of pain. Pondering the possible extinction of "the blacks" in 1869, French visitor Georges Clemenceau declared that "in this ruthless struggle for existence carried on by human society, those who are weaker physically, intellectually or morally must in the end yield to the stronger. The law is hard, but there is no use in rebelling."[61] In 1877, anthropologist Lewis Henry Morgan put a romantic spin on the well-worn theme of racial extinction, this time in reference to the nation's native peoples. Advanced society, Morgan insisted, owes its "present condition . . . to the struggles, the sufferings, the heroic exertions and the patient toil of our barbarous, and more remotely, of our savage ancestors. Their labors, their trials and their successes were part of the plan of the Supreme Intelligence to develop a barbarian out of a savage, and a civilized man out of this barbarian."[62] Such depictions of "heroic exertions and patient toil" conflated predestination, natural selection, race, and moral and intellectual progress in the encompassing frame of civilization.[63]

The mutable stages of evolutionary development, ranging from savagery to barbarism to civilization, echoed late nineteenth-century theories of individual development. The suppression of ancestral savages seemed to recapitulate the suppression of the savage within: only the painful suppression of individual habit, feeling, and desire could secure civility. In the decades following Emancipation, dietary guides, etiquette manuals, and the "lifecraft" texts of the scouting movement all imparted the importance of deliberate, willful labor in curbing surfeits of instinctual affect. The prevalence of such recommendations for "will training" indicates that the free self was not only perceived to be piloted from above—beholden to God's omniscient directives—but also by strange internal forces as well. The individual's messy emotional inheritance, Darwin noted with Victorian dismay, often leaked out despite his best deliberations: "Actions, which were at first voluntary, soon became habitual . . . and may then be performed even in opposition to the will."[64]

Put differently, the full self-possession that distinguished manly civility rested not only on the subjugation of nameless masses of "savage ancestors" but also on the forceful squelching of the myriad inner threats to the self's command. With the rise of psychoanalysis in the late nineteenth century, the role of suffering in the maintenance of willful self-governance gained new prominence. Freud advanced this view when proposing that psychic coherence entails the ego's surrender to the dictatorial superego. As a consequence of the civilizing process, he declared in 1912, "renunciation and suffering . . . cannot be avoided by the human race."[65] By the end of the nineteenth century, affluent intellectuals generally assumed that suffering must attend progress, whether barbarous peoples or the civilized ego paid the price.

Exaction of the price of progress depended on one's location on the evolutionary ladder: only some selves were allotted possession of individually consequential blood and tears. If not all suffering was thought to be equally valuable, this was because not all beings were persons and not all persons were similarly individuated—that is, equally endowed with the full complement of vulnerabilities that defined the proprietal self. In medical discussions of childbirth, for instance, the suffering endured by fragile white women was regularly contrasted with the deliveries of those whom civilization had not yet "deranged" (or, according to some commentators, "elevated").[66] Even the allegedly painless deliveries of women of "less-evolved" sensibilities could be recuperated into larger evolutionary narratives of difference. In an address before a New York medical society in 1899, an attending gynecologist from Mount Sinai Hospital noted that recent anthropological evidence countered the widespread conception that "savage races . . . give birth to their children while on the march, and . . . suffer less actual pain in the act of parturition." If "the Eskimos" make less of their pain, wrote Samuel L. Brickner, it is only because "their susceptibilities to pain of all kinds . . . [are] lower than those of the inhabitants of the more temperate zones."[67]

In short, even after Emancipation, sensibility to pain remained a specialized faculty, a sign of one's standing on an evolutionary scale. Physician Henry Bigelow invoked this comparative analytical frame when claiming in 1900 that "a Bushman or a Digger Indian" would hardly suffer vivisection more than "an intelligent dog."[68] The elaboration of this "great chain of feeling" (as historian Martin Pernick deftly titles it) expanded alongside the growth of medical professionalism, as legions of specialists developed their own novel methods for assessing and comparing bodily sensitivities.[69] In increasingly detailed taxonomies of difference, the invulnerability of deviants, primitives, and animals continued to provide the contrast against which the special selfhood of the sufferer emerged. "In our process of being civilized," concluded neurologist S. Weir Mitchell, "we have won, I suspect, intensified capacity to suffer. The savage does not feel pain as we do."[70]

Claims such as Mitchell's touched debates far outside the field of medicine; for as in the antebellum period, sensibility to pain was mapped onto the capacity for self-possession, with all of the consequences for political participation that capacity implied. Countless social battles—over the education of white women, the restriction of Chinese immigration, the legality of animal vivisection, and so forth—drew on evolutionary theories about the delicacy, hardiness, or insensibility of those to whom political autonomy might be granted in greater or lesser degree.[71] The editor of *Popular Science Monthly,* himself an advocate of higher education for women, described the evolutionary limitations of womanly anatomy in an 1874 editorial: "That woman has a sphere marked out by her organization . . . is as true as that the bird and the fish have spheres which are determined by their organic natures. . . . women may make transient diversions from the sphere of activity for which they are constituted, but they are nevertheless formed and designed for maternity, the care of children, and the affairs of domestic life."[72]

Similarly taking up the "'irrepressible" question of women's education, T. H. Huxley specified the connections between suffering and civic participation in 1896: "What social and political rights have women? What ought they be allowed, or not allowed, to do, be, and suffer?"[73] "The endowment of sensibility," summarized the former president of the Brooklyn Polytechnic when appearing before a Senate committee in 1900, "is the endowment of rights. Without sensibility there is no right. A being without sensibility can suffer no wrong."[74] Again, the relationship between sensibility and rights only increased in importance after 1870 as national citizenship gained new constitutional significance. Where once states had the ability to determine which beings were capable of full civic participation (evident in California's decision to exclude Chinese witnesses from testimony), the postwar amendments endowed the nation with that capacity. In this shifting legal context, the denial of sensibility effected the denial of full membership in the national body.

Thus, when Mills delivered his 1884 remarks on "Mental Over-Work among Public and Professional Men," he evoked much larger histories of religious, legal, economic, and medical thought. For the scientists of whom Mills spoke, as for Emerson's voluntaries, will proved essential to distinguishing meaningful suffering from helpless endurance. Whether assiduous work with a microscope or laborious collection of data would be considered heroic or degenerate, normal or pathological was bound to the privileged ability to *choose* suffering. Unlike those "historically burdened" by their bodies, the legacies of proprietal selfhood endowed scientists with bodily invisibility, and the capacity for full civic participation and testimonial credibility engendered by that invisibility.[75] And like the personified corporation whose voluntary servitude exempts it from the Thirteenth Amendment, the disembodied scientists of whom Mills spoke ("some minds") were both masters and servants, willing slaves who selected their own distinctive forms of bondage. Even as this voluntary self displayed definite privilege, it was a privilege based on obligation, on the compulsion to surrender one's person and its capacities. Elucidating how "science" came to address this indebted liberal subject—to occupy the role once allotted to God or nation in commanding deliberate suffering—is the task of the next chapter.

2

The Bonds of Science

Like the colonial theologians and regimental soldiers discussed in chapter 1, sacrificial scientists in the late nineteenth-century encountered a problem of exchange. Would suffering be compensated by new scientific knowledge? If so, just how readily ought one to embrace it? Might particularly painful, spectacular effort guarantee science's favor? As before, such questions gave practical form to the question of how one should live. And as before, answers to such ethical questions occasioned appraisal of the unseen force consummating the exchange. Where God or nation once bestowed value on the self's offerings, by the last quarter of the nineteenth century, science occupied a similar role.

The efflorescence of self-sacrifice in discussions of science after 1875 relied not only on a general preoccupation with voluntary suffering but also on an imagination of science as sovereign, immortal, and deserving of oblation. As discussed, the figure of the suffering scientist stemmed from the transformation of selfhood and property that attended the demise of racial slavery and the ascendance of industrial capitalism. However, it grew equally out of the increasing visibility and authority of science, which itself rested on two interrelated developments: the first primarily material, involving the growth and organization of institutions, standards, and networks; the second primarily symbolic, involving a transformation in vocabulary and customs of reference. Attending to both of these developments, in this chapter I recount how a staggering new array of buildings, practices, participants, methods, and findings came to be understood as aspects of the single, quasi-metaphysical entity *science*.[1] What relations existed between an emerging science and the postbellum voluntary self? Framing the connection between scientist and science as one of self-sacrifice opened scientists to not only the indeterminacy of the free self but also the ambiguous nature of the *gift*: that point at which "obligation and liberty intermingle."[2] This chapter illustrates the tangled and conflicting types of gift evident in scientists' sacrifices and addresses the particular roles allotted to science in framing those relationships.

THE ETHOS OF self-sacrifice evoked by late nineteenth-century scientists emphasized both voluntarism (the act must be intentional rather than accidental) and loss (it must be governed by something other than calculated self-interest). An 1895 essay by Edwin Emory Slosson exemplifies the importance of this principle of loss in scientists' usage. Writing for a popular New York newspaper, the young chemist from Wyoming argued that a common urge toward sacrifice unified seemingly disparate kinds of investigators:

> A biologist who wishes to study the life history of the tapeworm grows one in his own body; a physician ruins his health by experimenting on processes of digestion in his own stomach; a geographer risks his life to get a barometric reading; a bacteriologist inoculates himself only too successfully with a disease germ; a sanitarian, in order to test the effect of decomposed organic matter on the human system, drinks sewer water for a month; a chemist works for years on compounds so explosive that the careless touching of a few grains would kill him.[3]

Yet these cases were merely the most spectacular examples of "self-sacrifice." Even more theoretical or abstract scientific pursuits, Slosson continued, "cost someone years of toil or, perhaps, his life."

"Immense expenditure" stands at the center of this account of science. The rewards of scientific work do not "outweigh" individual sacrifices of life and limb. Suffering does not complete a grand theory of physical forces or unravel the mysteries of the human body. At best, Slosson proposed, self-sacrifice might help deduce the "fourth decimal" of some infinitesimal fact, the vaporous traces of a fundamentally unattainable truth. Indeed, voluntary submission to science's limitless movement is the scientist's defining attribute. "It may safely be said that if it were known that an important scientific discovery, say one which would fill a few lines in some large manual, could be made, but only at the cost of the life of the investigator, there would be no lack of volunteers." Far from appearing intrinsically calculating and acquisitive, the scientist here appears peculiarly prone to forfeiture. Science proceeds not through the investigator's cautious allocations of effort but through ostentatious giving—through what Slosson terms "self-immolation on the altar of science."

Slosson's self-immolating scientist reproduces the characteristic tensions of the "willing captive" discussed in chapter 1, including the ambiguities of possessive individualism inherent to the voluntarily suffering self (the biologist grows a tapeworm in his own body, the bacteriologist inoculates himself, and so on). At the same time, Slosson's remarks help introduce a second set of tensions, which similarly recur throughout the examples discussed in the four subsequent case studies: namely, the undecided character of the sacrifice. For late nineteenth-century scientists no less than for combatants in the war between the states, the sacrificial act must remain purposeful lest it be confused with simple self-destruction. The investigator who ate tapeworms just to eat them would not be giving something of himself for science; he would simply be deranged, degenerate, ignoble. The *reason* for activity was of primary importance in bestowing sacrificial meaning. Thus, the sacrificial gift entailed finding the appropriate balance of liberty and obligation: involuntary forfeiture—immolation by force rather than self-immolation—implied only the vile debasement of captivity; voluntary trade, such as contractual exchange, implied only the alienation of the market. The true self-sacrifice must be freely chosen but not *too* calculated. It must be willful—an intentional, reasonable offering—while remaining somewhat immoderate, uncompensated, "excessive in advance."[4]

As subsequent chapters will show, late nineteenth-century commentators diverged on their negotiation of this tension. Some suggested that science would amortize individual suffering; all that was needed was a reasoned assessment of the proper investment. Others argued that such judicious calculation was incompatible with the ethos of sacrifice. The scientist could never strategically barter suffering for knowledge since exchanging blood or labor in order to acquire a new fact annulled the meaning of the gift. For these proponents of self-sacrifice, an evident lack of compensatory reason is precisely what preserved the special status of the self's offering.

Some readers will recognize in such divergence echoes of a long-standing argument in social theory concerning the nature and function of the gift. In the most influential contribution to this debate, anthropologist Marcel Mauss argues that the free gift as such is illusory. In actuality, each gift perpetuates a complex system of reciprocity. Every item seemingly given ("food, women, children, property, talismans, land, labour services, priestly functions, ranks") possesses something of the giver, something that is "there for passing on, and for balancing accounts. Everything passes to and fro as if there were a constant exchange of spiritual matter."[5] There is no true, pure gift in Mauss's framework since the obligation to reciprocate persists: "exchange-through-gift" is the rule.[6] Challenging the presumptive freedom of the gift, his essay effectively troubles the notion of autonomous selfhood at the center of English liberalism. Supplanting ahistorical understandings of the free individual with a fundamentally social sense of personhood, Mauss emphasizes the obligations attendant on all actors, whose norms of conduct fluctuate with changing modes of production.

Yet critics point out that, in conflating gift and exchange, Mauss forecloses the question of uncompensated expenditure. "One cannot deny the phenomenon" of the exchanged gift, Jacques Derrida writes, but "the apparent, visible contradiction of these two values—gift and exchange—must be problematized."[7] Indeed, one might say that Mauss rejects the autonomous, voluntary self of

English liberalism, but only by elevating the utilitarian functioning of the total social system, its overall tendency toward reciprocity and solidarity. In contrast, Georges Bataille shifts the register of analysis from modes of production to modes of consumption, from a universal principle of stability and equilibrium to one of wasteful excess and purposeless destruction. Whereas Mauss viewed the "madly extravagant" destruction of the Northwest Indian potlatch as the "monstrous product" of an overarching principle of obligatory reciprocity, Bataille gives the potlatch fundamental paternity: immoderate squander is the real impetus of social activity.[8]

Rather than placing my reading of late nineteenth-century self-sacrifice in one or the other camp of this theoretical debate, I want to stress that the scientists discussed in subsequent chapters display similar equivocation on the nature of the gift. Moreover, their navigation of the tension between gift and exchange reveals the changing status of science, inscrutable recipient of the sacrifice. Whether *sacrifice* implied a kind of wholly excessive self-immolation or a form of stabilizing reciprocal exchange hinged on two related questions: the extent to which the suffering self was seen as wholly free (or in some way captive to or possessed by science) and whether science was seen as conforming to procedures comprehensible to its human subjects. Was science a person in the sense that its movements held significance, purpose, desire, aspiration?[9] Or was it more akin to a god, compelled by the bonds of covenant: *do ut des,* I give so that you may give?[10] The autonomy of science, in short, was as murky as that of the voluntary self.

We thus turn to an exploration of that for which one might choose to suffer. How did such a strikingly diverse set of endeavors (the biologist studying tapeworms, the geographer obtaining a barometric reading, the sanitarian testing the effects of decomposed organic matter) ever come to appear as aspects of a single body, one that might command sacrifice? By what processes were the leisurely investigations of a few curious gentlemen transformed into a dense transnational conglomeration of observers, experimenters, and theoreticians, an assemblage alternately personified or deified as "science"?

In many ways, this consequential transformation can be traced to the founding of local and national European academies in the seventeenth century: the Accadèmia del Cimento in Florence (1657), the Royal Society of London (1660), the Académie Royale des Sciences in Paris (1666), the Akademie der Wissenschaften in Berlin (1700), the Imperatorskaya Academiya nauk in St. Petersburg (1725), and so forth. Learned societies not only cultivated habits of assembly but also helped formalize and disseminate customs crucial to the eventual development of scientific institutions such as the preservation of journals, letters, and papers; the circulation of papers, specimens, and speakers; and the refinement of experimental methods. As historians have argued, the organization of learned societies marked an important watershed in the move from individual, localized investigation to collective, diasporic practices of inquiry.[11]

True scientific cosmopolitanism, however, did not arise until well into the nineteenth century, when a number of disparate events and processes converged. Improved postal networks, railways, and telegraphs allowed for enhanced communication across regions and nations, as did the standardization of units, categories, and instruments of observation and measurement. Increasingly complex and numerous interactions among individuals of varying skill and status hailing from diverse linguistic and cultural backgrounds wrought other changes as well. Successful collaboration between distant, heterogeneous observers required meticulous discipline, the careful formulation of behavior and speech. Over time, this discipline was formalized: qualifications for the certification of scientific competence were established, training procedures were implemented, reward systems were enhanced, and specialized congresses and commissions were organized. American investigators participated in these transnational developments in numerous ways, sharing specimens, articles, and informal correspondence; sending students abroad for training; hosting foreign visitors; and joining large-scale efforts such as observations of the 1874 and 1882 transits of Venus and the investigations conducted during the 1882–83 International Polar Year.[12]

American investigators further promoted this broad expansion and reorganization of scientific labor by reforming their domestic institutions. Among the most formidable of these changes was the transformation of American higher education that had been ongoing since the mid-nineteenth century. The passage of the Morrill Land Grant Act in 1862; the adoption of the elective system at Harvard, Yale, and other denominational colleges; and the establishment of new research universities such as Johns Hopkins (1876), Clark (1889), Stanford (1891), and Chicago (1891) created a radically different context for the production of knowledge. Graduate enrollment leapt from 198 in 1871 to 9,370 in the 1910s, and the number of scientific schools appended to academic institutions climbed from seventeen in 1870 to a height of seventy in 1873.[13] From these transformed colleges and universities emerged a growing middle class dedicated to the place of science in higher education. These Americans, whose everyday lives were increasingly filled with innovations such as the telephone (patented in 1876) and electricity (established with New York's Pearl Street station in 1882), were bombarded with claims that science had made these attention-grabbing applications possible. Broader interest in organized research soared accordingly, evident in the fanfare surrounding the opening of Edison's Menlo Park laboratory in 1876, the popularity of spectacles such as the Centennial Exposition in Philadelphia, and the growth of popular periodicals on science.[14]

Meanwhile, the establishment of federal agencies, such as an independent Department of Agriculture (1862, achieving cabinet status in 1889), the Bureau of American Ethnology (1879), and the National Bureau of Standards (1901), promoted a sense of solidarity among the various scientific investigators now dispersed around the country, as did the formation of new professional societies such as the American Chemical Society (1876), the American Society of Naturalists (1883), and the American Physical Society (1899). A series of new national journals such as *Science* (1883), the *Journal of the American Medical Association* (1883), and the *Physical Review* (1893) further unified investigators in disparate geographical, disciplinary, and institutional locations. The availability of large sums of capital upheld many of these changes as philanthropists enlarged endowments to libraries, museums, universities, and research agencies. In 1899 alone, donors gave $55 million to higher education, more than $5 million to libraries, and nearly $3 million to museums.[15] Through such developments, the "impersonal collectivity" acquired form and substance.

These institutional and economic transformations corresponded to an equally consequential conceptual shift, reflected in changing usage of the word *science*. As production of knowledge about the natural world garnered new financial and institutional clout, the word came to refer to an abstract, autonomous force, one with its own independent course of evolution. To be sure, science had been depicted as an identifiable thing before 1875. In an 1838 letter to Alexander Dallas Bache, for example, Joseph Henry spoke of his desire to "advance the cause" of science, a desire realized in the founding of the American Association for the Advancement of Science in 1848. A generation earlier, the American Philosophical Society employed similar language, noting that its members were "animated by a love of science."[16] Yet these usages were unusual in American letters before the Civil War.[17] In the antebellum period, *science* more typically evoked a type of mental discipline, a faculty or quality peculiar to the human mind. This conception of science lingered late into the century. As the editor of *Popular Science Monthly* wrote in 1872, science is not regarded as "applying to this or that class of objects" but "as being, in fact, a method of the mind, a quality or character of knowledge upon all subjects which we can think or know."[18]

By 1875, however, references to science as an attribute of the human mind had largely evaporated. No longer implying a form of cerebral discipline, the word instead summoned forth an independent body with its own intrinsic imperatives. Physicist T. C. Mendenhall made this shift plain in an 1890 address, when he delineated the public's "obligations to science and her votaries." "Thanks to science," he declared, "time and space are practically annihilated; night is turned into day; social life is almost revolutionized and scores of things which only a few years ago would have been pronounced impossible, are being accomplished daily."[19] As Mendenhall's words suggest, by 1890 *science* implied an exacting and captivating agent, one said to merit deference from "her" votaries.[20] Although the boundaries of science remained flexible and contentious (it might or might not include grammar, theology, or history, for example), after 1875 *science* almost invariably appeared in American writings as a disembodied entity with independent command.

It might be said that science was coming to be experienced as an "imagined community," in the words of political theorist Benedict Anderson.[21] Like the nations of which Anderson speaks, this imagined community had four chief attributes. First, science was *imagined* in that most of its members would never know one another. Scattered by discipline, institutional affiliation, status, and geographical location, one would never see or know all of the colleagues toiling away in distant field sites, observation stations, laboratories, or classrooms. Second, it was *limited* in that even the most expansive concepts of science presumed boundaries beyond which resided other domains of living: art, politics, medicine, industry, religion, and so on. Again, exactly which of these overlapping domains might be considered part of science remained subject to continual debate (for example, is medicine scientific?); but in each case, "science" was delimited. Third, science was *sovereign* in that, like the nation, it acquired authority as the role of divine governance receded. Indeed, what distinguished science from earlier traditions, historian Andrew Cunningham argues, "is that Natural Philosophy was an enterprise . . . about God." In contrast, science "is an enterprise which (virtually by definition) is *not* about God. God, His existence and attributes are taken to be *irrelevant* to science and the practising scientist."[22] Fourth, science was a *community*: as Anderson writes of the nation, despite the "actual inequality and exploitation" apparent in its institutions, science was "always conceived as a deep, horizontal comradeship," a fraternity between members (as Lorraine Daston puts it) "equal in their anonymity."[23] Ultimately, Anderson proposes, it is this fictive sense of fraternity that enables so many people not merely to kill for the sake of the imagined community but also "willingly to die" in its name.[24]

The imagination of science as a sovereign subject allowed for a reorganization of time—a submerging of the individual's present life for the sake of the envisioned future of the collective body. Like the soldier willing to die for the nation or the disciple willing to die for God, the scientist's deferral to a delayed destiny secured the suspension of immediate interest necessary for the continuation of science. Max Weber offered perhaps the most famous explication of this deferral in his 1918 speech, "Science as a Vocation." Here Weber argued that because science's continual expansion denies any prospect of culmination, the scientist *must* subjugate himself to the future growth of knowledge. To accentuate his vision of science as constantly expanding, Weber distinguishes it from art. While artists might share with scientists certain attributes (like single-minded devotion), science differs from all artistic endeavors in its denial of restitution:

> Scientific work is chained to the course of progress; whereas in the realm of art there is no progress in the same sense. . . . A work of art which is genuine 'fulfillment' is never surpassed; it will never be antiquated. . . . In science, each of us knows that what he has accomplished will be antiquated in ten, twenty, fifty years. That is the fate to which science is subjected; it is the very *meaning* of scientific work, to which it is devoted in a quite specific sense, as compared with other spheres of culture for which in general the same holds. Every scientific 'fulfillment' raises new 'questions'; it *asks* to be 'surpassed' and outdated. Whoever wishes to serve science has to resign himself to this fact.[25]

Artists might toil and suffer in the process of their work. Artists might valorize self-sacrifice. But unlike scientists, Weber insists, artists *can* attain fulfillment. In his view, the "very meaning" of science rejects such fulfillment. Without a willingness to defer satisfaction indefinitely, to "ask to be surpassed and outdated," scientific work could not proceed. The scientist's resignation to a life of forfeiture is the basis of science's existence.

It would be a mistake to take Weber's 1918 comments on voluntary subjection as an accurate assessment of the intrinsic nature of science (or, for that matter, of art). As noted, the idea that science as a discrete thing "chained to the course of progress" was itself an artifact of the nineteenth century. Weber's comments concern us here not as evidence of the timeless character of science but as a way to consider the norms of scientific selfhood being configured at that time. Imagining science as eternal positioned subjects of ethical reflection as members of another, immortal body. "Let it be remembered," wrote Cincinnati astronomer Ormsby Mitchel, "that the astronomer has ever lived, and never dies. The sentinel upon the watchtower is relieved from duty, but another takes his place, and the vigil is unbroken."[26]

Indeed, a repositioning of the subjects of science necessarily attended the growth of supranational networks of investigation in the nineteenth century. As Daston, Galison, and Levine have argued, novel forms of collaboration among far-flung investigators entailed novel expectations for knowing subjects, epistemological ideals better suited to the types of transnational practices coming into being at midcentury. In place of a focus on distinctive, trustworthy individuals, which characterized inquiry in the seventeenth and eighteenth centuries, there arose a call for observers "unmarked by nationality, by sensory dullness or acuity, by training or tradition; by quirky apparatus, by colourful writing style, or by any other idiosyncracy that might interfere with the communication, comparison, and accumulation of results." Drawing primarily on English, French, and German sources, Daston and others demonstrate how the nineteenth-century investigator's individual persona was to be submerged into the communal body of science. "L'art c'est moi," as physiologist Claude Bernard declared; "la science, c'est nous.' "[27]

American investigators joined this transatlantic revision of norms, expanding and elaborating their engagement as the upheaval of Civil War slowly settled. In their depictions, the relationship between the self and the increasingly potent figure of science was clearly one of subordination. "Man," declared *Scientific American* in 1890, is but "the ward of science."[28] Man, echoed the *Pedagogical Seminary* in 1901, while not "an organ in the body of science" is at least "a cell with functions truly vital."[29] Whether positioned as ward to guardian, cell to body, or proselyte to deity, the characteristic forms of subjection attributed to the scientist changed in tandem with the characteristic forms of command imputed to science. Each vision of science's governance effected a distinct vision of self-governance and vice versa.

The advent of the word *scientist* suggests the mutually constitutive relationship between science and its subjects. The word entered the lexicon for the first time in March 1834, introduced by English polymath William Whewell. Reviewing a treatise on the physical sciences, Whewell lamented the "disintegration" of investigative unity: the "mathematician turns away from the chemist; the chemist from the naturalist; the mathematician, left to himself, divides himself into a pure mathematician and a mixed mathematician." To combat the increasing fragmentation of practices of inquiry, Whewell proposed the umbrella term *scientist* as an appropriately "general term by which these gentlemen could describe themselves."[30] His neologism gained currency slowly, however, and in the United States displaced more common terms such as *experimental philosopher* or *naturalist* only late in the century, as *science* came to appear as an agent in its own right.[31]

The slight but meaningful distinction between the recent term *scientist* and the older *man of science* attests to the arrival of a novel and still unsettled subject of ethical reflection. Like *man of the cloth,* the phrase *man of science* implies an individual who exists independently of the collective, an identifiable agent who still requires a preposition ("of") to join the larger body. With the dissemination of the word *scientist*, this emphasis on the independence of the willful human recedes; he belongs to it, is possessed by it. Science's status as the subject in whose name action takes place is accordingly enlarged. In place of "man," "science" itself now demands, revolutionizes, determines; the scientist, the person defined by science, merely responds.

A flexible ethic of self-sacrifice helped mediate the emergent relationship between the imagined body of science and the freshly minted figure of the scientist. As discussed, the phrase *self-sacrifice* came into English in the early nineteenth century, at the same moment that property-in-the-person superceded

the ownership of dispensable wealth as the defining condition of civic membership. Self-sacrifice's increasing prominence after 1875 can be said to reflect increasing strain on liberal ideals of autonomous selfhood and consensual
contract in changing historical circumstances—including increasingly professionalized and specialized practices of inquiry. As Simon Newcomb concluded
in 1904, high among the "common elements and common principles" that bind
"the seemingly unending subdivision of knowledge" is the idea that science "offers an even nobler field for the exercise of heroic qualities than . . . that of bat-
tle."[32] Like other Gilded Age Americans seeking bulwarks against rampant
greed, scientists labored to differentiate the alienable from the inalienable,
seeking in science a way to protect the sacred "me" from the profane, salable
"mine."[33]

The tensions evident in these efforts—tensions that express larger social
contradictions—were given poignancy and immediacy by the matter of voluntary suffering. Would the amputation of one's fingers or toes serve to advance
knowledge, or were these efforts ultimately futile and destructive? Was nonpro-
ductive expenditure—the deliberate squander of useful things—precisely the
point, in that it lifted both science and scientists beyond the debasing calculations of the market? As I will discuss in subsequent chapters, divergent answers
to these questions were possible. Some writers described scientific work as a
contractual relationship, in which parts of the self (stomach, hands, thought)
were temporarily bartered based on assumed returns of equal or greater value.
Others described a process of deliberate disowning, a principle of forfeiture designed to distance themselves from the accumulative mindset of "the masses."
Confusion and conflict over the meanings of sacrifice reflected the contradictory pressures of American life in an era of unfettered capitalism.

Again, such debates over sacrifice were not the only avenues available for
framing the relationship between the scientist and science. The point is not
that self-sacrifice was ubiquitous in late nineteenth-century science but that
the spread of self-sacrifice at this time suggests an emergent form of ethical reasoning, new modes of working on the self. The triumph of liberalism left the status of the gift oddly uncertain. In the age of the contract, the reason of and for
sacrifice was ever in question, for scientists no less than for others.

3

Purists

In a rousing address delivered to incoming students in 1875, Harvard professor of chemistry and mineralogy Josiah Cooke readied his charges for the painful life that lay before them. To grapple with nature, Cooke insisted, one must be prepared to "labor long in the dark before the day begins to dawn." The toil of scientific work, however, wouldn't end with mere deprivation of slumber. Ideally, one should strive to exceed in zeal that "student who would cut off his right hand rather than be guilty of a conscious untruth."[1] Reprinted in *Popular Science Monthly*, Cooke's interweaving of devotion, truth, and suffering presaged themes that would soon reverberate through the writings of university-based researchers. Drawing on well-worn images of ignorance and wildness, Cooke's Harvard colleague Charles Gross described in 1894 how the investigator's singular "zeal for truth" drove him into "the dark forest of the unknown," a "love of patient, disinterested, conscientious labor" overwhelming any fear of danger.[2] Similar allusions permeate the essays of psychologist G. Stanley Hall, who regularly proposed that the "great edifice" of science had been built by men "content to spend laborious lives, to become laboratory hermits, to explore with great hardship, labor and risk and even in inhospitable lands, to deny themselves."[3] Beginning in the 1870s and extending into the first years of the twentieth century, commentators such as Cooke, Gross, and Hall explicitly tied an embrace of suffering to a special love of truth. Arguing that the true aims of scientific inquiry had become confused with the vulgar, materialistic ends of practical application and financial advancement, such reformers promoted "pure science" as an ennobling antidote to the masses' preoccupations with utility, profit, and ease.

Such promotion was hardly inevitable. In 1835, Alexis de Tocqueville famously declared that "the pure desire to know" was utterly absent in America and confessed his skepticism that a passion for truth might ever arise in a democratic nation, particularly one so "addicted" to practicality.[4] Although Alexander Dallas

Bache and some members of the Lazzaroni began promoting truth for truth's sake as early as the 1850s, these calls remained exceptional until later in the century. Instead, unflattering comparisons between U.S. science and the "higher interests" of continental Europe remained standard fare in the writings of foreign visitors for nearly forty years.[5] Yet by the mid-1880s, advocates of pure science could be found in governmental organizations, learned societies, colleges, universities, museums, and private funding agencies. In freshly revamped ecclesiastical colleges and new research universities, advocates of "science for science's sake" held increasing influence. The generation of scientists born after 1840 saw support for pure science materializing in new institutions, such as Columbia's School of Pure Science (1890), and in new endowments for research, such as Elizabeth Thompson's large donation to the American Association for the Advancement of Science in 1873.[6]

Historians have long discussed late nineteenth-century investigators' infatuation with pure science and have described the "emotional resonance" that themes of purity held for the first generation of university-based scientists. These scholars have illustrated how proponents of pure research, under the influence of German academic fashion, expressed disdain for commercialism, exhibited conviction in the moral significance of labor, and promoted inquiry as a path to professional and national advancement. Their work has shown not only that advocates of pure science steered the course of developing research universities but also that these universities, in turn, altered the shape and scope of scientific research.[7] This chapter complements and extends that work by revealing the centrality of voluntary suffering to these developments.

Pure science, its advocates proposed, was distinguished by its refusal of utility and practicality, by its removal of men from the more "sordid things of the world." Unlike applied science, which generates "comfort, pleasure and happiness," the "modern spirit of pure science . . . elevates man's ideals."[8] As we will see, this elevation was effected by demonstrations of hardship: the purist's embodied displays of deprivation testified to his emancipation from material concerns. Through voluntary suffering, proponents of pure science incorporated endeavors increasingly marked by institutional specialization and fragmentation, claiming diverse and disparate practices as part of the single, demanding body of science.

This deliberate embrace of toil, poverty, and fasting, and the understandings of purity it enacted, reveal the characteristic paradoxes and exclusions of the sacrificial self. For while the will to suffer helped to demonstrate the purity of the investigator's devotion to science, the fleshly evidence of this suffering was conditioned by existing systems of privilege. Only the fully self-possessed scientist might indulge the call to sacrifice.

PURITY IS AN ancient concept, and the coupling of purity and suffering promoted by late nineteenth-century university reformers had deep roots. The term *Puritan*, first used as a term of opprobrium by early critics, evokes the sect's torturous efforts to regain the all-encompassing religiosity of the first-century church.[9] As Puritanism took hold in the New World, its creed of moral elevation through toil and renunciation grew into the splintering branches of American Protestantism. The dominant motifs of Christ's sacrifice, including its valorization of transcendence through suffering, remained central to New England theological traditions through the nineteenth century.[10]

Most late nineteenth-century proponents of pure science imbibed while still in childhood this brew of suffering, moral uplift, and worldly labor. G. Stanley Hall was taught to affiliate suffering with spiritual zeal as a member of his deeply religious western Massachusetts family. When his maternal aunt died in adolescence, Hall's grandfather stood with his hand on his daughter's coffin and declared it "the happiest day" of his life since he "felt the love of God so shed abroad in his heart."[11] Other advocates of pure science shared similar backgrounds. Of the men who entered scientific fields between 1861 and 1876, more than 27 percent had fathers employed in ministry, a greater percentage than in any other occupation.[12] The eminent botanist John Coulter was born in China to missionary parents, while geologist Thomas C. Chamberlin was the son of the well-known Methodist minister John Chamberlin. Hall, Ira Remsen, Henry Rowland, and other vociferous proponents of pure science all studied for the ministry before choosing scientific careers. While they moved away from religious service, they maintained Protestant ideals of purification through worldly labor.[13]

Such ancient religious ideals of purity took particular shape in the unsteady political context of post-Reconstruction America, as those contesting the boundaries of the nation and the rights of individual citizens tapped into inherited Christian ideals. In the period of pure science's ascendance, no political and economic controversy evoked the language of purity and pollution more than the issue of Chinese immigration discussed in chapter 1. Debates over anti-Chinese exclusion laws focused on the threatened purity of American civilization, a purity understood as simultaneously moral, intellectual, and physiological. In an influential 1862 treatise on Chinese immigration, California physician Arthur B. Stout laid out the themes of defilement and infestation that would become increasingly prevalent over in the 1870s and 1880s. The "great West," Stout proclaimed,

> is overwhelmed with a Chinese immigration. Once permitted it must be forever endured. The work of degeneration once commenced, its progress must pursue its insidious and empoisoning influence, not for a few years, but for centuries to come. The legislation now enacted is less for our own that for generations which, in the future, by their purity shall bless, or in their degeneration shall curse, their ancestral stock.[14]

Such calls for legislative prophylaxis were soon heeded: the special taxes on foreign miners in place since the 1850s were followed by more overt prohibitions and restrictions, including the Page Law of 1875 and the first of several Chinese Exclusion Acts in 1882. In the 1870s and 1880s, references to "purity" invariably conjured ongoing dispute over race and civic participation.[15]

Indeed, the entanglement of moral, physiological, economic, and racial norms recurs in the writings of proponents of pure science. These entanglements are helpfully pronounced in physicist Henry Augustus Rowland's famous 1883 "Plea for Pure Science" before the American Association for the Advancement of Science, an address that one historian aptly dubs the "ne plus ultra of pure science rhetoric" in the nineteenth century.[16] Widely cited by university reformers, Rowland's speech displays the role of voluntary suffering in emerging definitions of pure science, including the familiar exclusions and paradoxes which that suffering entailed. The remarks at once reflected the period's racialized discussions of purity and civilization and invigorated the links between these discussions and the imagined body of science.

As a child, Rowland seemed destined for a life in the clergy: he was the son and grandson of Congregational ministers and the alleged descendent of Jonathan Edwards. Yet by the age of twenty, Rowland abandoned a future in the ministry and declared it his "duty and vocation to be an investigator in science." "I intend to devote myself hereafter entirely to *science*," he wrote to his mother in 1868. "If she gives me wealth I will receive it as coming from a friend but if not I will not murmur."[17] Particularly talented in optics, magnetism, and electricity, Rowland pursued his interests first at Wooster University, then at Rensselaer Polytechnic Institute, and finally as a faculty member of the new Johns Hopkins University. Described by a former colleague as "supercharged with new ideas," he quickly became one of the foremost investigators of his generation, known in both the United States and Europe for his work with spectral diffraction gratings. In the summer of 1883, at the height of his professional renown, Rowland traveled to Minneapolis to the annual meeting of the American Association for the Advancement of Science, where he delivered the vice-presidential address of the association's physics section.[18]

The address sparked discussion across the country. Shortly after Rowland's speech, physicist and engineer William A. Anthony told the association that Rowland hardly fulfilled his own stated ideal of pure science. As Anthony pointed out, Rowland's position as an industrial consultant and holder of several patents made him a rather hypocritical proponent of the nobility of nonutilitarian research. Anthony also drew attention to the elitism of Rowland's position: "Few *can* devote their lives to work that promises no return except for the satisfaction of adding to the sum of human knowledge. Very few have both the means and the inclination to do this. Most of us are dependent upon salaries, and a salary binds us to service which, unfortunately, does not, in this country, usually mean scientific research."[19] Others, however, heard in Rowland's critique of profit-oriented research echoes of their own growing dissatisfaction with the commercialism of American science. In a letter to the president of Johns Hopkins, physicist John Trowbridge described the popularity of the speech's message among his colleagues at Harvard. The Harvard men, Trowbridge noted, were themselves too timid to lash out at "mercantile professional chaps" as Rowland had done.[20] Other approving listeners urged publication of the address; and "A Plea for Pure Science" was soon reprinted in *Science*, the *Journal of the Franklin Institute*, *Popular Science Monthly*, and an array of other periodicals.[21]

A close reading of Rowland's text illuminates the place of suffering in defining pure science. The address begins by differentiating *pure science*—a term conflated with *science* throughout the speech—from its "applications," conveniences such as telegraphs and electric lights. To make the somewhat hazy distinction between pure science and its applications more clear, Rowland distinguishes the types of men engaged in each activity. A "people like the Chinese," he suggests, have satisfied themselves with the applications of science, using gunpowder without analyzing its chemical properties. By failing to pursue further research, the Chinese have "fallen behind in the progress of the world; and we now regard this oldest and most numerous of nations as only barbarians." If "Americans" hoped to resist a similar slide toward barbarity, they must display the "moral courage" necessary to obtain new knowledge of scientific theory. Rowland here assumes that the nation's evolutionary position is mutable: it might advance through courageous work or regress through sloth and inattention—a position he shared with other university reformers of the time.[22]

Moral courage is required, he continues, since pure science necessarily entails hardships, particularly financial impoverishment. Unlike the applications of science, which offer immediate economic rewards, the "higher pursuit" of scientific study attracted no monetary benefits. The researcher's courageous disavowal of material concerns therefore becomes a prerequisite for pure science. Echoing the declaration of poverty he made to his mother fifteen years earlier, Rowland insists that the pure scientist must not be guided by acquisitiveness: "If our aim in life is wealth, let us honestly engage in commercial pursuits, and compete with others for its profession; but if we choose a life which we consider higher, let us live up to it, taking wealth or poverty as it may chance to come to us." The stigmatizing connotations of poverty are thus elided through the matter of choice: the atavistic barbarian haplessly endures material impoverishment, but the "great and unselfish workers" of the past *elect* this condition. The exemplary "poor man," Michael Faraday, represents the "few, the very few, who, in spite of all difficulties, have kept their eyes fixed on the goal."[23] The embrace of poverty demonstrates the choice itself: Faraday's renunciation of material goods and his pauper's death testify to the purity of his devotion to truth.

Scientists such as Rowland were hardly unique among late nineteenth-century professionals in trading lamentations about salaries with sympathetic colleagues.[24] Yet unlike commentary on law or medicine, disregard for financial affairs here acted to define pure science as distinct from alternate modes of investigation. For Rowland, poverty constitutes the boundaries of the endeavor: a man made wealthy by his research would, by definition, no longer be pure.

Poverty, however, is not the only hardship said to demand the pure scientist's moral courage. According to Rowland, a tolerance for social isolation is equally important for those abandoning applied or commercial endeavors. Because pure science requires unencumbered travel in the furthest reaches of the unknown, investigators must be willing to deny their inherent conviviality. Since "Man is a gregarious animal, and depends very much, for his happiness, on the sympathy of those around him," the pure scientist's ability to curtail these animal desires is crucial to his success. While stopping short of declaring the scientist's need for chastity, Rowland's speech stresses the importance of solitude. In the United States, pure science necessitates a particular tolerance for loneliness, he informed the crowd gathered in Minneapolis, since Americans are uncommonly hostile to nonprofitable endeavors. In such a materialistic nation, the pure scientist "must be prepared to be looked down upon." As with the endurance of financial hardship, the investigator's suppression of his innate sociability becomes synonymous with pure science itself. In Rowland's terms, "pure science is the pioneer who must not hover about cities and civilized countries, but must strike into unknown forests, and climb the hitherto inaccessible mountains which lead to and command a view of the promised land."[25]

So intense are the myriad hardships of pure science, he suggests, that only those who share a "love of nature" will be moved to participate; the nature of the self determines the nature of the science and vice versa. Because the work of pure science is so difficult, so lacking in financial reward, and so opposed to man's "gregarious" nature, most will be driven away from its pursuit. The true scientist, however, possesses ardor sufficient to compel the embrace of toil, poverty, and loneliness. Unlike applied scientists, pure scientists are not motivated by the common aims of wealth, fame, or societal improvement but by "pure love." "We must live such lives of pure devotion to our science," Rowland maintains, "that all shall see that we ask for money, not that we may live in indolent ease at the expense of charity, but that we may work for that which has advanced and will advance the world more than any other subject, both intellectually and physically."[26] The purity of the science, in other words, lies not in the object of inquiry but in the inner affect of the scientist—in the moral character of his desire.[27]

These affections, essentially internal and invisible, are made evident by the scientist's persistence in the face of suffering. Devotion impels him through the

vale of solitude and poverty; at the same time, endurance of these hardships reveals the purity of his veneration. Insisting on the necessity of isolation and
poverty, Rowland implies that these difficult external conditions affirm the scientist's inner commitment: "if the spirit is there, it will show itself in spite of
circumstances."[28] The scientificity of the man—that is, the purity of his devotion
to truth—is never so vivid as when placed against the backdrop of loneliness
and destitution. In the words of Rowland's famous ancestor, Jonathan Edwards,
"true virtue never appears so lovely, as when it is most oppressed."[29]

Rowland's speech does not suggest that devotion alone can support pure
science. Alongside glowing descriptions of Faraday's noble endurance of poverty,
Rowland lobbies for greater public provision for pure scientists and their research within the collective space of the university. Unlike the novelist, the
poet, and the musician, Rowland complains, the scientist lacks any external in-
centive to work. While a "chivalric spirit" might encourage rare men such as
Faraday to devote their lives to the study of nature, to "cultivate this highest
class of men in science, we must open a career for them worthy of their efforts."
In time, Rowland maintains, the presence of full-time pure scientists within the
university would transmit the importance of this work to future generations.
"Young men, looking forward into the world for something to do, see before
them this high and noble life, and they see that there is something more honorable than the accumulation of wealth. They are thus led to devote their lives to
similar pursuits, and they honor the professor who has drawn them to something higher than they might otherwise have aspired to reach."[30]

Rowland thus identifies the university as a special preserve for those men
who share a "love of nature," that elite group with the "ability" and "taste for
higher pursuits." So great is the emphasis on this special ability that it appears
the more "the masses" despised an endeavor, the more pure it became.[31] The
elite devotees of science—whom Rowland insists are neither Chinese nor "red"
men nor, apparently, women of any color—are distinguished by their willingness to endure whatever obstacles the pursuit of science might entail.[32]

Recalling earlier theological debates over the nature of free will, whether
these "better feelings" are entirely inborn or whether men might elevate themselves through perseverance and hard work remains an unanswered question in
Rowland's remarks. While emphasizing the civilizing effects of toil, he also insists that "it is a fact in nature, that no democracy can change, that men are *not*
equal,—that some have brains, and some hands; and no idle talk about equality
can ever subvert the order of the universe."[33] The point to reiterate here, however, is that endurance of hardship, renunciation of wealth, and disavowal of
companionship helped to demonstrate the scientist's purity. Whereas telegraphs and electric lights indicate applied science, pure science is signaled by
the poverty and loneliness of its practitioners. For a science characterized by
the scientist's inner devotion, such ordeals provide crucial testimony to the

integrity of internal commitment, whether inborn or acquired. Palpable suffering renders visible the self's moral character.

To be sure, suffering was not the only route to truth found in narratives of scientific advancement. Rowland himself distinguished these paths when accepting a medal from the American Association for the Advancement of Science, segregating results stemming from "long and persistent endeavor" from those obtained by "happy accident." Willful labor, he reminded his audience, must not be confused with fortuity.[34] Nor did all scientists glorify voluntary suffering. As Rowland idealized the scientist's groveling pursuit of an ever-distant truth, other self-described scientists were pathologizing avowed commitment to continually deferred satisfaction. In emerging studies of masochism, displays of willful debasement troubled rather than enlarged the subject's perceived capacity for rationality.[35]

Yet the "Plea for Pure Science" amplified themes flowing through the writings of university-based advocates of pure science in the 1880s and 1890s.[36] Like Rowland, these advocates generally held salaried positions in new research universities, had been reared in pious Protestant families, and considered ministry before choosing scientific careers. And like Rowland, they tied the purity of science to the scientist's willingness to suffer. Pure scientists' unique "spirit of inquiry," it was routinely suggested, propelled them through the inevitable hardships of the scientific endeavor.[37] The scientist's extraordinary devotion to truth was said to inspire an uncommon tolerance for "toil, solitude and penury."[38] Like the stigmata of a tortured saint, the scientist's tribulations, an embrace of "unremitting toil," manifested an ineffable, interior devotion to truth.[39] And in each case, pure science was defined by this inner affect. As University of Michigan physicist Henry Carhart argued, passionate zeal for truth distinguished the pure scientist: "It is a gross libel on scientific men to assert that the chief end[s] . . . are narrowly utilitarian or intensely practical. If worldly success were the only reward awaiting the scientific investigator, but few branches of science would be fortunate enough to find their votaries. The taste for scientific research is a passion which finds its gratification in the truth it seeks."[40]

As in Rowland's "Plea," these discussions display a persistent ambivalence about whether this inner purity was inherent and fixed or whether it might be acquired and cultivated through deliberate labor. They also waffle on the attribution of causal agency: in some instances, the desires of the scientist direct science's unfolding; in others, the demands of science direct the scientist's desires. Science, Hall wrote in 1894, "transfigures work so that men come to love nothing so well as difficulties to be overcome." Like the domineering "Venus in Furs" of Leopold von Sacher-Masoch's famous 1870 novel, in this instance scientific truth holds the dominant role in the relationship, rewarding only "those who truly love and wait upon her."[41] Writing in 1900, Charles Sanders Pierce similarly emphasized the scientist's servility but instead placed agency with the "man of science" moved by his "deep impression of the majesty of truth . . . that to which, sooner or later, every knee must bow."[42] In pure science as in other domains of life after Reconstruction, the free, volitional self was constantly in tension with the demands and imperatives of larger abstractions. Like the Christian, the pure scientist displayed his will most by submitting to a higher power.

The scientist's devoted subservience emerged in tandem with this power, the object for which one chose to suffer: a truth that negates completion and satisfaction. In the opening decades of the nineteenth century, "truth" was hardly so mercurial. Instead, it was generally represented as a comforting edifice of universal knowledge, stable and timeless. The scholar's task was to study this unchanging edifice, to know and represent its fine contours. Yet with the shift from truth to objectivity as the dominant epistemological ideal after midcentury came a corresponding sense of the ephemerality and incompleteness of human knowledge. The arrival of the principles of Darwinism, the abandonment of the Newtonian emission theory of light, and other radical reorientations of thought challenged any notion of truth as comprehensive and lasting. Everywhere there seemed new evidence for what Lorraine Daston has called "the Heraclitean lesson of *panta rhei*": nothing in stasis, everything in tumultuous upheaval.[43] As Ernst Mach noted in 1872, "attempts to hold fast to the beautiful moment through textbooks have always been futile. One gradually accustoms oneself [to the fact] that science is incomplete, mutable."[44] Because truth was a permanently receding object, the pure scientist was accordingly imagined as a doting follower, Sisyphean in his relentless toil. Echoing the capitalist imperative of privation in the name of a future, always-deferred reward, the seeker of truth similarly affirmed renunciation in the name of a distant, indeed unattainable goal.[45] Pure scientists praised one another for engaging in the restless pursuit of an ephemeral ideal—"like the chase after the foot of the rainbow, which ever moves onward as it is pursued."[46]

Emphasis on scientists' servility was of particular consequence to pure science, a field whose most palpable attribute was the flesh of its practitioners. As an endeavor identified by a lack of material productivity—as that "for which there is no reward except the glory thereof"—pure science could only be demonstrated by its subjects themselves.[47] Just as useful inventions such as telegraphs and electric lights provided the tangible incarnation of applied science, so the fatigued, impoverished body of the researcher came to manifest pure science. The pure scientist's body suggested his internal predilections, the integrity of his deep love for truth. While working out a difficult and forceful idea at Johns Hopkins, for example, eminent mathematician James Joseph Sylvester sat up "a good part of the night with his feet in warm water to check the rush of blood to his head."[48] Absorbed in the course of his research, Harvard's Charles Gross often neglected to take meals, a habit thought to have hastened his fatal illness. Perennially concerned that he was faltering in the pace of his research, Gross's

Harvard colleague Herbert Levi Osgood frequently rose from bed at two o'clock in the morning to continue his work. Watching Osgood, his colleagues recalled, "one could almost actually see the element of will pitilessly driving a poor body to the limit of its power."[49] The fact that prominent scientists were rarely seen outside the laboratory was taken as further testimony to the purity of their devotion to truth. Socially inept researchers were lauded for their ability to devote themselves even "more assiduously, not only days but nights," to their scientific work."[50] Invoked in each instance as the natural premise for such behaviors, the pure scientist's moral integrity was constituted through his laborious suffering; manifest misery signaled an invisible, interior devotion to truth. Impoverishment, fatigue, social isolation, and other displays of deprivation substantiated the ardor of the researcher's devotion and hence the intensity of his proximity to pure science.

In the eyes of its proponents, the university offered a particularly important location in which to live out this devotion. This was not because the university was free from commercial interests. Rather, the university was distinguished as a site for pure science by a relative lack of stricture. At a time when Darwinism still drew ire and censorship in denominational colleges, the absence of external compulsions (from administrators, clergy, or trustees) emerged as the crucial difference between the university and the conventional American college. The "true distinction between Collegiate and University education," said the president of Tulane University in an 1884 address, is that the "former is training, the latter emancipation and liberty."[51] "If it is the truth that makes us free," Johns Hopkins professor Basil Gildersleeve wrote in 1893, "it is freedom that opens the way to truth."[52] To effectively "fan the burning coals of that enthusiasm which is absolutely untainted by any sordid motives into the intensest glow," Hermann von Holst argued at the first convocation of the University of Chicago, students must be granted a "large measure of liberty," not only with respect to "the What, but also as to the How, When, and How Much."[53] The first president of Johns Hopkins, Daniel Coit Gilman, staunchly defended the ideal of individual freedom. In response to unsolicited advice, he once angrily declared his unwillingness to place "fetters put upon a professor in any department of science."[54]

Within the university, defined as that emancipated space where external strictures on curriculum were relaxed or abandoned, the researcher might work himself to the limits of his strength. The absence of collegiate rules and regulations translated into increased self-scrutiny. Discipline ceased to be a matter of external coercion (an authority's assessment of the rate, amount, and content of work performed) and became a matter of inner compulsion. Supposedly inspired and sustained by a deep love for truth, scientists were compelled to maintain "eternal vigilance," a "constant watching of themselves."[55] Themes of strict self-governance were repeated throughout the celebrated translation of

Friedrich Paulsen's *The German Universities and University Study,* adopted by American educators as a blueprint for their own reforms.[56] In the absence of "external compulsion," Paulsen wrote, "the more imperative is the duty of self-control. Whoever confounds freedom with license, misunderstands its meaning; it is given to the individual not that he may do as he pleases, but that he may learn to govern himself."[57] Or as historian von Holst stated succinctly in 1893, "incessant work is the price of liberty."[58] At Johns Hopkins, historian Owen Hannaway has argued, Gilman's major influence was to transfer the force of control from "the general curriculum and the external conduct of the university (where great liberality was granted) to the intimate atmosphere of the laboratory and seminar," in which students and faculty became ever more responsible for their own discipline.[59]

Such calls for exacting self-discipline and exhaustive effort helped craft a remarkably consistent and singular vision of science, one all the more noteworthy in light of the transformation of practices of inquiry then ongoing. As we have seen, the passage of the Morrill Land Grant Act in 1862, the adoption of elective systems at denominational colleges, and the establishment of universities made possible by vast industrial fortunes altered the material conditions of organized research. In the context of this institutional growth and reorganization, descriptions of suffering played a unifying role, bringing a sense of interconnection and purpose to intellectual work increasingly fragmented and specialized. Within the walls of a single university, researchers were segregated by department, field, division, and school. Even as investigation grew ever more diffuse and disunified—divided among universities, philanthropic institutes, industrial laboratories, experimental stations, and observatories—the word *science* increasingly referred to a single, autonomous entity that demanded toil, privation, and penury. A letter from mineralogist Marshman Edward Wadsworth, penned in response to Rowland's "Plea," expressed hope in the deep, underlying unity of seemingly distinct realms of knowledge and practice, a unity revealed in "true scientific work." Wadsworth tentatively suggested that Rowland's remarks applied equally to both "natural" and "physical" sciences, fields knit together by a common "duty" to truth.[60] Institutional cleavage, in other words, might be healed by the shared wounds of pure scientific work.[61]

Even as voluntary suffering provided a sense of common purpose for investigators such as Rowland and Wadsworth, it created schisms along lines of sex and race, distancing advocates of pure science from other researchers in the last quarter of the nineteenth century. As discussed in chapter 1, the capacity for self-sacrifice presumed a self socially endowed with personal autonomy and bodily integrity. Black and female bodies, which had been demarcated by and through systems of dependency, commodification, dispossession, and exchange, could scarcely call on ethics of voluntary suffering to the same extent as Rowland and his colleagues. While devotion to truth might bring a man like

Osgood to the limits of his power, the will to suffer was structured by existing positions of privilege.

For white women and black women and men with interest in scientific research, agency tended to take other forms: namely, the pursuit of projects with clear practical or profit-minded goals in areas such as medicine or agriculture. Astronomer Maria Mitchell (1818–89) discovered as much in 1876, when she obtained responses to a questionnaire sent to white women investigators across the country. That such women were more likely to pursue "those sciences which touch life and health" stemmed only in part from humanitarian sentiments. Of equal consequence was the pressing matter of opportunity: postgraduate training and employment in physics, astronomy, or mathematics were simply more difficult to attain than work in charitable organizations, medical missions, or agricultural research.[62] Barriers to white women's participation in more abstract fields were layered, multiple, and perpetuated even by the most vociferous advocates of "higher" research. Despite Rowland's pleas for pure science, for example, he was one of the most strident opponents to women's participation at his own university, Johns Hopkins. Although women were admitted to the Hopkins medical school in 1893, they were not officially admitted to the university's graduate school until 1907, thirty-one years after its founding. Even then, individual faculty members reserved the right to refuse women entry to their laboratories and classrooms.[63]

Some agitators sought to draw on norms of purity and sacrifice when promoting new research opportunities for women. For example, one contributor to *Popular Science Monthly* proposed that any science advanced by a woman was "pure" by definition.[64] Another argued that enlisting "the enthusiasm, the self-sacrifice and vitality of women in the cause of science" would not only advance knowledge but also enliven women and the men who lived with them. Where women were "complaining and sentimental, they will grow cheerful and wise"; where there had been "restless longings" would come instead "continuous and pains-taking labor."[65] Mitchell herself conjured existing conventions of feminine self-sacrifice in her 1875 presidential address before the Association for the Advancement of Women. Women's disregard for self, she argued, made them exceptionally suited for the work of science: "In my younger days, when I was pained by the half-educated, loose, and inaccurate ways which we [women] all had, I used to say, 'How much women need exact science,' but since I have known some workers in science who were not always true to the teachings of nature, who have loved self more than science, I have now said, 'How much science needs women.' "[66] Basing resistance to institutional barriers on women's unsullied capacity for self-sacrifice, however, implied a complicated bargain: access to science then depended on one's special willingness to suffer. Moreover, none of these advocates for women appear to have considered the racial and socioeconomic specificity of this idealized femininity—further reproducing the familiar exclusions of the sacrificial self.

Black men and women were also systematically barred from graduate schools, salaried employment, and professional organizations and similarly migrated toward more available opportunities in biological, medical, pharmaceutical, and agricultural fields. The myriad legal, economic, physical, and social obstacles to black participation in pure science structured period debates over industrial education. Booker T. Washington alluded to these obstacles when asserting that black education "'should be so directed that the greatest proportion of the mental strength of the masses will be brought to bear upon the everyday practical things of life, upon something that is needed to be done, and something which they will be permitted to do in the community in which they reside."[67] His words suggest the conflict between ideals of unfettered inquiry and the concrete realities of systemic racism in the closing years of the century. Sandwiched between inequities in secondary educational funding and inadequate employment opportunities for postgraduate scientists, most black colleges chose to support the model of vocational education promoted by Washington, influencing the scope of scientific research for years to come.[68]

In the context of ongoing struggles over access to the material resources necessary for organized research, recurrent emphasis on the nobility and civility of suffering for pure science played a particular role. As Rowland's "Plea" makes clear, it is the directed, deliberate character of the scientist's exertions, the volition emphasized throughout the address, that differentiates Anglo-Saxon men of science from the effeminate, degenerate, pathological sufferings understood to be helplessly, irrationally endured by women, barbarians, or masochists. The scientist *chose* his suffering and, by choosing to suffer, both demonstrated and reinforced his elevation. Yet while a privileged legacy of bodily ownership secured one's ability to suffer for science, that position of agency and control depended on the continual exhibition of its characteristic vulnerability.

Displays of suffering—relentless toil, social isolation, fasting, penury—thus became the language in which pure researchers spoke of their commitment to a constantly retreating higher truth, and research universities provided space for this display. The interpenetration of pain, truth, and devotion found in pure science is particularly evident in the university laboratory of Ira Remsen (1846–1927), where students enacted the tensions of scientific suffering. Remsen's chemical laboratory at Johns Hopkins became a key location for transmitting the disciplines of purity and exemplifies the power of an ethic of suffering in a specific community.[69] (Clark University, one of only two all-postgraduate institutions in the country at the time, was described by some late nineteenth-century commentators as "more pure" than Hopkins. Yet Clark never achieved equivalent renown and hence never exerted such a continuing and considerable effect on the shape of science education in the United States.)[70]

Unlike his Hopkins colleague Rowland (who crowed of neglecting students for his research), Remsen was known for a strong commitment to instructing future generations of scientists, and he placed special emphasis on the value of pure science. In Remsen's view, the pursuit of science is "nothing but" the expression of a desire to know, which "we have implanted in us."[71] Truly pure science must thus be distinguished from applications and all other research conducted with profit in mind. "It is the province of science to investigate, to discover, to know, to furnish the material or the knowledge that is to be applied," he argued, "but it is not its province to apply."[72] Like Rowland, he filled his descriptions of curiosity-driven science with tales of the toil and poverty of great scientists of the past. This elevation of the impoverished seeker of truth made an impression on Remsen's students, who recalled themselves at Hopkins as "noble young men dedicated to learning and poverty."[73]

For those students not yet dedicated to the pursuit of pure science, Remsen reiterated the moral significance of science pursued without practical use in mind, often describing his contempt for hands "sull[ied]" with industry.[74] Best known for his role in the synthesis of benzoic sulfinide (saccharin), he was somewhat ambivalent on the matter of commercialism.[75] Although himself a consultant to General Chemical, Standard Oil, and numerous other industrial and governmental bodies, he actively dissuaded his students from pursuing such appointments. When one student suggested that impoverished universities might seek industrial sponsorship for their laboratories, Remsen "responded that he could think of no worse fate for the university than such an invasion." Despite his consultations for large businesses, he was, as Owen Hannaway has argued, a "pious worshipper at the shrine of pure science" and a disciple who actively sought to convert his students.[76]

To his students, Remsen also imparted the virtues of hard work. He always insisted that a certain amount of arduous and tedious experimentation was necessary for research, particularly in chemistry. Remsen frequently cited his mentor, Justus von Liebig, on this point: "Any idea that stimulates men to work, excites the perceptive faculties and brings perseverance is a gain for science, for it is work which leads to discoveries." Like many of his peers, he was fond of describing the "exceedingly severe" examinations of German students and how lacking American universities were in the ability to turn out a similarly "full-grown scientific man."[77]

Furthermore, Remsen considered rigorous work as ceaseless as the expansion of knowledge itself. No matter how much perseverance and thoroughness a student might exhibit, there was always more work to be done: discovery is necessarily ceaseless. "Every so-called 'complete' investigation," he reminded his students, "is surrounded by question marks." The universe, he suggested, held "a power 'that passeth all understanding.' "[78] In knowledge's limitlessness, Remsen found an easy affinity with his early religious training. Like many fellow advocates of pure science, he had undergone a strict Protestant upbringing. After his mother's death in his early childhood, Remsen was sent to a small village to live with his stern maternal great-grandmother and his great-grandfather, a pastor in the Dutch Reformed Church. In his adult life, Remsen retained his childhood piety and expressed no conflict between his religious and scientific beliefs, even clipping devotional articles to glue inside his professional notebooks.[79] In his words, "the *ultimate* of both science and religion are *infinites* . . . something that gives meaning to all that passes, and yet eludes apprehension; something whose possession is the final good, and yet beyond all reach, something which is the ultimate ideal, and the hopeless quest."[80] Like the quest for religious truth, the search for knowledge was always receding; the work required of the researcher was correspondingly ceaseless.

While Remsen emphasized ceaseless work, poverty, and the desire for knowledge in his discussions of pure science, he differed from many of his peers in an important way. Unlike Rowland, who dramatically emphasized the individual researcher's lonely ascent of the scientific mountaintop, Remsen always stressed the communal aspects of pure science in both his professional publications and his daily interactions with students. He refused to lead a reclusive life and encouraged students to see themselves (to cite the title of an 1912 address) as "all members one of another."[81] Adamant that students should not work in isolated chambers, Remsen arranged laboratories at both Johns Hopkins and other universities around a large central room to encourage contact among students. When he helped plan a new chemical laboratory at the University of Chicago, for example, he designed rooms for students to gather separately from their instructors, remarking that "students learn more from each other than from their teachers."[82]

To foster continuous interaction among these disciples of pure science, Remsen carefully delineated scientific from nonscientific spaces, starting with the racially and sexually marked bodies long kept to the vestibules of civic society. Neither black students nor white women students were admitted to Remsen's laboratory; and even those students allowed into the space were, for the larger part of each day, cut off from the rest of the university community.[83] The careful purification of the laboratory extended to supposedly defiling objects. When the Spanish-American War broke out, Remsen refused to allow certain "base" newspapers into the laboratory, although everyone was admittedly interested in the contents of the articles. Within the exclusive space of the laboratory, students were taught to be fastidious about its condition, not only about the neatness and cleanliness of the equipment and surfaces but also about behaviors. Smoking was not allowed in Remsen's laboratory, and no student felt comfortable arriving in shirt sleeves.[84] Many recalled its church-like atmosphere, and students emerged from Hopkins imagining themselves to be "high priests . . . bound to the truth"—a truth perpetually just out of grasp.[85]

It is here that the significance of voluntary suffering in pure science begins to emerge, for in Remsen's laboratory students drove themselves to incessant work. Although Remsen insisted that tedious exertion was necessary to scientific advancement and that American students were not nearly as driven as their impressive German counterparts, he did not force students to toil in his laboratory. On the contrary, he was known to remark disapprovingly on the increasing "pressure for more work in all departments," and the fact that such "hurry and worry" left little time for reflection. When a group of students came to him to complain about the amount of work they had been assigned in another class, Remsen quickly took steps to alleviate their obligations in his mandatory chemistry lab. He maintained strict hours in the laboratory, shutting off the gas at five o'clock sharp to discourage lingering.[86]

Remsen's laboratory, then, displayed not dictatorial control but the social power of a norm of science marked by the capacity and desire for voluntary suffering. Although he encouraged his students to maintain regular hours, they instead enacted the assumption that the purity of one's commitment was demonstrated in the vigor of one's exertion. Becoming model voluntary sufferers, students learned to monitor their own habits of investigation, "determined to produce something more in research to please the Old Master."[87] The general impression among laboratory students was that "nobody ever cuts Remsen."[88] Sharing stories of peers who pursued truth with exceptional diligence, students breathed further life into the valorization of what Nietzsche has called "self-chosen torture."[89] One Hopkins alum recalled a legendary student who "kept an all-night's vigil in one of the laboratories, like Don Quixote watching his armour." Two other students were found to have broken into a biological laboratory on campus to continue their work after hours. Understanding their efforts to be generated by devotion to an infinite truth, and believing that the existence of their devotion distinguished them from everyone outside the laboratory, students came to determine their own work discipline.[90] An invisible, ineffable commitment was made palpable and veritable through the medium of the flesh. Custom had created within the community practices of deliberate privation that further elevated pure science as an object worthy of sacrifice.

Remsen's students went on to found chemical laboratories in universities across the United States, disseminating his pedagogical style and the Hopkins's model of self-governance.[91] The university, in turn, shaped future understandings of science. By 1895, the retiring president of the Philosophical Society of Washington could claim that Americans had done little to advance applied science; investigators' chief contributions, wrote G. Brown Goode, had been "in pure science rather than in the application of science."[92] By the turn of the century, some American researchers lamented that applied science had come to require cultural and financial reinforcement. Chemical expert Ellen Swallow Richards argued in 1911 that the basic merits of the "useful dollar" had to be forced upon a powerful "aristocracy of learning," which had grown resistant to all practical work.[93] Despite Tocqueville's earlier predictions to the contrary, "science for science's sake" had acquired currency in the United States, a currency incarnated in the suffering bodies of its proponents.

At stake in the suffering evident among late nineteenth-century proponents of pure science, then, was a telling feature of their civility: a willing subjection to habits of deprivation and toil. These displays of willful subjection were condemned to repetition, since purity ran the risk of being contaminated at every interval. As evident in the incessant toil of Remsen's students and the excessive mutilations described by Slosson, establishing the purity of one's science was not a singular act but a routine, methodical, and recurring activity. Truth was now ever elusive and the search for truth never ending. Maintaining purity therefore demanded a constant and uncompromising self-governance, staking proponents of pure science to ceaseless labors of love.

4

Explorers

As with pure science, a certain lack of utility defines Arctic polar exploration. Explorers themselves are quick to point out that, to reach the pole, is to reach a "fruitless" place.[1] Devoid of the gems, minerals, and spices of Africa and the Americas, the frozen polar sea offers no tangible reward. In the words of one historian of the Arctic, the pole "might be regarded as the most useless piece of real estate on earth."[2] Of course, there were furs to be had in some parts of the Arctic, oil in others, and even the pole itself could turn a profit: the names of America's two most famous North Pole explorers, Frederick A. Cook and Robert E. Peary, would eventually be used to hock everything from Bibles to footwear. Yet turn-of-the-century polar explorers emphasized the Arctic's manifest lack of practical usefulness. "Nowhere else," concluded one Norwegian explorer in 1911, "have we won our way more slowly, nowhere else has every new step cost us so much trouble, so many privations and sufferings, and certainly nowhere have the resulting discoveries promised fewer material advantages."[3]

This recurring emphasis on the uncompensated character of explorers' privations highlights the dilemma of contractual exchange underlying the ethos of sacrifice in the late nineteenth century and the tensions between voluntarism and coercion at the heart of the contract. To what end were polar explorations undertaken, if not for evident "material advantages"? How did American explorers, and the middle-class reading public to which they were bound, frame the relation between knowledge of the Arctic and individual suffering?

This chapter examines the explorers' ethos of willing sacrifice, tracing familiar patterns of exclusion and ambiguity in their accounts of voluntary privation. While depictions of suffering have suffused narratives of exploration for millennia, the conjunction of science and exploration in the late nineteenth century reformulated understandings of individual sacrifice, tying American polar explorers to the larger traditions of possessive individualism I have already discussed. Focusing on statements circulated by and about explorers Peary and Cook, I show how their access to volitional suffering—and the testimonial credibility that this access helped engender—depended on existing forms of racial, economic, and sexual privilege. As before, this access was not absolute or fixed; and fractional levels of credibility and vulnerability were afforded variously to bodies understood as Eskimo, Negro, or white, female or male. Moreover, the privilege to suffer for Arctic science was caught in the familiar paradoxes of liberal selfhood: the tensions between liberty and compulsion, self-possession and self-forfeiture. Even as Cook and Peary described themselves as self-reliant pioneers, their descriptions of volitional suffering illuminate their entanglement in new forms of dependency and obligation.

The chapter treats only U.S. discussions of exploration; other nations affixed their own normative repertoires to the Arctic. (In Sweden, for instance, turn-of-the-century polar explorers were often likened to ancient Vikings.)[4] American explorers were profoundly shaped by events and processes unfolding transnationally, and I discuss the most relevant of these developments, showing how international changes in practices of polar exploration encountered the revisions of selfhood, science, and sacrifice ongoing in other spheres of American life, which in turn furthered new principles of reasonable suffering.

IN LIGHT OF explorers' repeated stress on the disutility of Arctic exploration, explaining their willingness to return to the death-dealing pole provides a challenge. One historian concludes that so much hardship endured for so little obvious reward provides "vivid testimony to the irrational element in exploration."[5] Explorers themselves have referred to this irrational element as "Arctic fever," a "malady" for which "there is no cure but to put the patient on ice."[6] Others describe the compulsion to reach the pole as a "drug," a "spell," or a strange "instinct" akin to a sexual drive.[7] Most frequently, turn-of-the-century American explorers portray their motivations as essentially religious, not a drug but a quest. In their accounts, the ice becomes an "Arctic Cathedral," the explorers become "disciples," the pole a part of the "Heavenly Kingdom."[8]

These religious images might be said to emerge from the nature of the pole itself. For unlike the stick of wood discovered by Winnie-the-Pooh during Christopher Robin's "expotition," the pole is not a perceptible landmark on the earth's surface.[9] It is instead an invisible ideal, an imagined mathematical point in the center of an ice-littered ocean. The "boreal unknown" can be approached only haltingly; physical movement must be paired with repeated mental corrections, adjustments, and computations.[10] Although at first glance, the pole seems eminently reachable (unlike the receding truth pursued so diligently by "pure scientists"), it exists only in another, immaterial space. One polar voyager recalled standing around nervously in 1909 as the expedition's commander lay on his stomach calculating sextant observations on a piece of tissue paper. After weeks spent sledging across the restless, semifrozen polar sea, the end of the journey was revealed not by a mark on the ice but by some apparition within the slim paper's smudged figures.[11] Another explorer reported a similar stuttering advance on the pole in 1908. Determining his position to be 89 degrees, 59 minutes, 46 seconds, he realized the pole was "in sight." He "advanced the fourteen seconds, made supplementary observations and prepared to stay long enough to permit a double round of observations."[12] The pole moved into the explorer's field of vision only after he collected further data and meditated on the relationship between those data and the laws of spheroidal trigonometry. At the globe's northernmost point, geographical exploration—the quintessential scene of western man's confrontation with new physical worlds—becomes an extended meditation on the intangible, the abstract, the meta-physical. In the words of one commentator, the North Pole represents "human striving for what is approachable but never fully attainable."[13]

The seemingly timeless affinity between the search for the pole and "human striving" for the infinite, however, should not blind us to the particularities of the religious metaphors used to describe late nineteenth- and early twentieth-century American expeditions to the Arctic. Unlike the genteel devotion advocated by university-based pure scientists, turn-of-the-century polar exploration appears to rely on a bloody, lethal obsession. The quest for new knowledge entailed not merely penury and solitude but a ride with death itself. Consider one widely reprinted 1909 poem, a paean to controversial Arctic explorer Robert Peary. The second stanza of Elsa Barker's "The Frozen Grail" reads:

> To conquer the world must man renounce the world?
> These have renounced it. Had ye only faith
> Ye might move mountains, said the Nazarene.
> Why, these have faith to move the zones of man
> Out to the point where All and Nothing meet.
> They catch the bit of Death between their teeth
> In one wild dash to trample the unknown
> And leap the gates of knowledge. They have dared
> Even to defy the sentinel that guards
> The doors of the forbidden—dared to hurl
> Their breathing bodies after the Ideal,
> That like the Heavenly Kingdom must be taken
> Only by violence.[14]

For Barker, religion and polar exploration do not share pacific human striving but instead hurling, bodily violence. This assumption, as we will see, came fairly late to the field of Arctic exploration and was tied to broader transformations in practices of science and norms of manliness. Our guides through this discussion will be the American explorers Cook and Peary.

The feud between them is well known. In the autumn of 1909, fifty-three-year-old Robert E. Peary sent word to the Associated Press in New York City that he and his four-man party had reached the North Pole on 6 April 1909—the first time in history. Peary's triumphant declaration, however, arrived just five days after another American, forty-four-year-old Frederick A. Cook, declared that he and two assistants had reached the pole in April 1908.[15] Peary instantly denounced Cook as a liar and a fraud. Cook, initially reserved about Peary's claim, eventually branded Peary as an adulterer, murderer, *and* liar.[16] The men's bitter dispute over polar priority soon engaged much of the country, pitting self-described "Cook-Americans" against legions of "Pearyites."[17] Countless articles were printed with titles such as "Dr. Frederick A. Cook—Faker"; countless photographs appeared such as one of an Eskimo woman and child, allegedly abandoned by Peary in Greenland, captioned, "Polar Tragedy—A Deserted Child of the Sultan of the North and Its Mother."[18] Ignited by these inflammatory publications and fanned by the enthusiastic participation of hundreds of thousands of opinionated citizens, the controversy occupied the attention of the National Academy of Sciences, the U.S. Congress, and a number of international geographic organizations.[19] For two full years the dispute permeated news reports on both sides of the Atlantic.

My discussion of Cook and Peary will neither attempt to recount the controversy in detail nor resolve it once and for all. Both tasks have been taken up thoroughly by previous scholars.[20] Rather than focus on the explorers' acrimonious differences, I want to draw attention to the ethic of science they shared, expressed so cogently by Elsa Barker: that knowledge of the Arctic required human suffering. The historical specificity of this ethos can be made visible by considering the early years of American exploration of the Arctic.

At the time of Cook's and Peary's separate attempts to reach the North Pole in the winter of 1908–9, the drive for the pole was widely represented as one of the great efforts of western civilization. In the inimitable words of Peary, the pole was a "mystery which has engaged the best thought and interest of some of the best men of the most vigorous and enlightened nations of the world for more than three centuries."[21] While most late nineteenth- and early twentieth-century commentators also affirmed that reaching the pole was the "climax" of a centuries-old quest, others stretched civilization's fascination with the pole even further into antiquity.[22] For these writers, ambition to reach the pole had burned in the hearts of men since "the adventurous galley of Pytheas of Massila, about 330 B.C., first brought back tales of a frozen ocean in the North."[23] In this case, attainment of the pole was said to be the culmination of an ancient human desire.

In point of fact, the search for the geographic North Pole was a relatively recent obsession. During the many centuries that European explorers scoured the globe's northern waters for expeditious passages to Asia, the pole itself remained an item of little concern. It was, after all, a rather intangible goal. Reaching the northernmost end of the globe's axis of rotation became an objective for explorers only in the early nineteenth century, augured by the polar quest that opens Mary Shelley's classic 1816 novel, *Frankenstein*. Within two years of the creation of Shelley's fictional Captain Walton, the sudden appearance of large quantities of floating ice off Greenland suggested to real mariners the prospect of open water at higher latitudes.[24] Within a generation, the pole had replaced commercial passage as the great destiny of maritime men.

This new fascination with reaching the pole reflected a number of broader developments: the increasing importance of European nationalisms and naval competitiveness; the enhanced organization of geographical societies; intensified interest in ethnological, botanical, meteorological, and geological investigations in distant locations; changing Victorian conceptions of manly duty and honor; and the rise of imperialism.[25] Britain launched the craze for polar exploration with expeditions commanded by Captain David Buchan and Captain John Ross in 1818.[26] On the heels of these attempts, explorers from Canada, Scandinavia, Austria-Hungary, Italy, and Russia also headed for the imagined top of the world.

Americans entered this frenzied exploration only after 1850.[27] The turning point in U.S. exploration of the north, as for the expansion of Arctic exploration more generally, was the loss of the British expedition led by Sir John Franklin. Franklin's vessels, the *Erebus* and the *Terror*, were observed by two whaling ships just a few months after departing from England in May 1845; two long years then elapsed without any further contact with the expedition. In the decade following 1847, more than forty search parties from Britain and abroad scoured the Arctic in search of the missing ships and their passengers.[28]

Partly through the efforts of Lady Franklin, who pleaded with the governments of all "civilized nations" to aid in the search for her husband, finding the expedition swelled with the sense of righteous purpose.[29] Franklin and his crew were represented as the embodiments of manly virtue, and the effort to locate them became a "noble enterprise":

> It is not merely scientific research and geographical discovery that are at present occupying the attention of the commanders of the vessels sent out; the lives of human beings are at stake, and above all, the lives of men who have nobly periled everything in the cause of national—nay, of universal progress and knowledge;—of men who have evinced on this and other expeditions the most dauntless bravery that men can evince.[30]

This swelling sense of moral purpose, paired with an apparently sordid fascination with the fate of the individual men, at last spurred American explorers to join the international rescue mission. John M. Clayton, writing on behalf of the U.S. president, told Lady Franklin that the country would be happy to support the search, as "the name of John Franklin has been endeared by his heroic virtues, and the suffering and sacrifices which he has encountered for the benefit of mankind."[31] In 1850, the *Advance* and the *Rescue*, two ships provisioned for three years by the American merchant Henry Grinnell, departed for the north.

The intertwined refrains of noble motivation and terrible sacrifice that circulated in discussions of the Franklin search expeditions set the tone for a generation of American exploration in the Arctic.[32] Even after explorers gave up the search for Franklin and assumed other aims in the Arctic, voyages were still framed as lofty Christian undertakings. America's foremost Arctic explorers—Elisha Kent Kane, Frederick Schwatka, Isaac Hayes, and Charles Francis Hall (each of whom had each been initiated into Arctic exploration through the Franklin searches)—continued to stress both the grim suffering and pure aspirations of their work. Disdainful of commercial motives, men who strove toward the frozen pole were said to possess inspirations more "philanthropic" than "mercenary."[33] Obviously, the rhetoric of philanthropic selflessness obscured the expeditions' ongoing interest in lucrative discoveries such as mineral deposits or whaling bays. Yet for this first generation of American polar explorers, the dominant tropes remained those articulated in the search for the missing Franklin expedition: pure devotion to a higher cause.

Explorers' accounts of severe suffering and hardship in the Arctic intensified the rhetoric of pure devotion. Starvation, frostbite, scurvy, maddening loneliness, polar bears, falls through the ice—such were the myriad perils said to exist for polar travelers. According to popular narratives of exploration, men who were willing to explore the "dismal realms" of the far north shared one "great element of distinguishing greatness, of which the explorers of more genial and inviting climes were destitute. Their investigations were made entirely without the prospect of rich reward."[34] The nobility of polar exploration, in other words, emerged through the suffering it entailed. Richly illustrated, best-selling testimonies such as Kane's *The United States Grinnell Expedition in Search of Sir John Franklin* (1854) and *Arctic Explorations: The Second Grinnell Expedition in Search of Sir John Franklin in the Years 1853, '54, '55* (1856) taught the American reading public to associate Arctic exploration with loss, danger, and death.[35] A willingness to undertake risk and hardship served to highlight the apparent selflessness and nobility of the endeavor.

Notably, the figure of science played little role in these popular accounts, despite the fact that the first generation of American expeditions included men trained in meteorology, ethnology, natural history, and astronomy. In the first three decades of U.S. exploration in the north, the few commentators who considered the relations between science and exploration portrayed both "science" and "men of science" as far removed from the frozen, sunless Arctic. The purity of Arctic exploration was rarely connected to the observation and recording of what commentators termed "natural phenomena." To explorers, pure interests in the north simply implied noncommercial motivations, not the investigation of nature.

Indeed, the physical hardships of Arctic work were generally said to obstruct investigations of nature. Being a man of science in the mid-nineteenth century implied the cultivation of self-renunciation, rigor, and studiousness—all traits coded as "manly" in the Victorian age. Yet this genteel view of investigation tended to emphasize mental and spiritual discipline, not physical endurance. Such moral, intellectual, and above all sedentary norms of inquiry could hardly be squared with contemporaneous norms of exploration: limbs shriveled from frostbite, eyes blinded by snow glare, and bellies distended from malnutrition. *Science,* defined at midcentury as a certain refinement of mind, seemed a far cry from the rotting whale meat and howling dogs populating Arctic narratives.[36]

The perceived opposition between science and Arctic exploration is exemplified by the final expedition of Charles Francis Hall (born 1821), a Cincinnati newspaper publisher who, at the age of thirty-eight, sold his business to fund a trip to the pole. Although one of Hall's central objectives was to investigate new bays suitable for whaling, he consistently represented his mission as sustained by noncommercial aims.[37] He sprinkled his pre-travel journals with quiet pleas to God for the success of his northern quest, and in one 1870 letter to the Senate Committee on Foreign Relations he stressed that "neither glory nor money has caused me to devote my very life and soul to Arctic exploration" but a concern with the pole for its own sake.[38] After Hall's death at Thank God Harbor in 1871, his eulogists highlighted the intensity of his devotion. Colleagues recalled that "so thoroughly had he identified himself with his work, that his feelings in regard to it began to assume the form of a religious enthusiasm." Hall trusted, "with the religious earnestness and faith of a sincere enthusiast, that he would finally reach the object of his devotion."[39]

While Hall emphasized the righteousness of his motives, he did not present the purity of his devotion as entailing a particular devotion to science. Certainly the voyage had what we now would consider a significant scientific component: the expedition dedicated three men (surgeon Emil Bessels, astronomer R.W.D. Bryan, and meteorologist Frederick Meyer) to the observation and investigation of natural phenomena. Moreover, when President Ulysses S. Grant signed a bill in 1870 providing money for the operation, he did so on the provision that investigations be performed on board the vessel "in accordance with the advice of the National Academy of Sciences."[40] Yet Hall never interpreted these details as reflective of the mission's aim. Although individual investigators, such as Chief of Corps Bessels, might be praised for their "devot[ion] to science," the endeavor as a whole was never presented as a scientific venture.[41] Joseph Henry stressed this fact when delineating the academy's instructions to Hall in 1871, stating gruffly that the expedition "is not of a scientific character."[42] In Henry's view, the aims and methods of polar exploration were antithetical to the aims and methods of science, for the simple fact that "men of the proper scientific acquirements" would hesitate to join "an enterprise which must necessarily be attended with much privation, and in which, in a measure, science must be subordinate."[43] Men of science, Henry suggested, would be repelled by the prospect of spending several months in frozen darkness, thousands of miles from the nearest library. For Henry and his peers, the mental discipline known as science had little in common with the perilous world of Arctic exploration.

But the remarks of Joseph Henry, a man born in the eighteenth century, represented a dying vision. No longer was science conceived as a genteel manner of thought; it was now seen as an entity with its own intrinsic needs. Attending this entity was a new figure: the scientist. By the last quarter of the nineteenth century, Henry's gentlemanly investigators had been supplanted by this new figure, the incarnation of emerging professional standards. The rise of the scientist, one outcome of the transnational transformation of inquiry occurring after midcentury, restructured both norms and practices of Arctic exploration.

Curiously, as the work of research was professionalized, both experimenters and theorists began to liken their work to the physically and emotionally taxing work of exploration. Exploration, in fact, became a metaphor for all scientific investigation. For physicist Henry Rowland, as we have seen, the figure of the virile explorer emblematized science as a whole: "pure science *is* the pioneer who must not hover about cities and civilized countries, but must strike into unknown forests, and climb the hitherto inaccessible mountains which lead to and command a view of the promised land."[44] As Rowland and his university-based colleagues emphasized, to strike out into such inaccessible terrain required certain forms of endurance: an ability to suffer social isolation, financial impoverishment, even physical pain. Promoting the particularly difficult (and eminently fundable) nature of their project, professionalizing scientists borrowed the metaphors of geographical exploration. In the process, they incorporated themes of dirty toil and physical suffering long held to be antithetical to the work of gentlemanly natural philosophers. Once demarcated from exploration in the matter of privation and hardship, science slowly came to share an emphasis on messy, grueling labor.

While physicists, chemists, psychologists, and historians were coupling their work with exploration, explorers themselves were seeking to align their endeavors with the aims and methods of science. The efforts of German-born naval lieutenant Karl Weyprecht exemplify this shift. Returning from an 1872–74 North Pole expedition, Weyprecht chastised the Austrian Royal Geographical Society and Britain's Royal Geographical Society for considering polar exploration "merely as a sort of international steeple-chase, which is primarily to confer honour upon this flag or the other."[45] In place of national boasts about minute gains in latitude, Weyprecht proposed collaboration among countries to enhance the observation and investigation of natural phenomena. The poles, he insisted, must be considered resources for the contemplation of more general and encompassing laws of nature. To this end, competitive impulses for territorial acquisition must be subdued in favor of a coordinated search for facts "profoundly linked to phenomena close at home."[46] Weyprecht explicated the principles of "scientific" polar exploration in a landmark 1875 paper, which insisted that geographical discovery in the Arctic has value "in as much as it prepares the way for scientific exploration as such" and that "for science the Geographical Pole does not have a greater value than any other point situated in high latitudes."[47] Such principles reframed the North Pole. No longer considered an independent object to be won and forgotten, the as-yet-unreached pole was now imagined as fundamentally linked to both the South Pole and to more temperate regions. Problems of meteorology, terrestrial magnetism, and astronomy, according to Weyprecht and his peers, were global in nature.[48] As the poles were integrated into more general scientific problems, so, too, were polar explorers incorporated into broader scientific communities.

Weyprecht's critical intervention fell on receptive ears. In October 1879, an International Polar Conference gathered in Hamburg to discuss how to bring science more fully into exploration. Eleven nations ultimately pledged support for the construction of a set of circumpolar observation stations under the direction of an International Polar Commission headed by Georg von Neumayer, a geophysicist with connections to Alexander von Humboldt, Justus von Liebig, and other prominent investigators of the age. During the first year of this collaboration, the First International Polar Year (1882–83), fourteen stations were put into operation around Arctic Circle. All told, more than seven hundred men participated in the multinational effort—one of the monumental transnational organizations discussed in previous chapters. American officials, eager to elevate the nation's status in such international affairs, ordered the construction of two stations: one at Point Barrow, Alaska, headed by Lieutenant P. Henry Ray, and one at Lady Franklin Bay, headed by Adolphus Washington Greely.[49]

Thus, as professionalizing scientists advanced an association between science and exploration that emphasized suffering, they were met by explorers (who had long emphasized their own sacrifices) who were conscientiously striving to render their activities scientific. These two groups, in turn, were part of larger middle-class communities who newly assumed the violent costs of progress. Although middle-class Americans debated whether knowledge merited the expenditure of human life, they assumed a certain system of exchange: Arctic knowledge must be purchased with human blood. The assumption of this system of exchange—and its distance from the views of science and exploration evident in the case of Charles Francis Hall—is illustrated by popular responses to two expeditions of the early 1880s, one led by Lieutenant Greely, the other by George Washington De Long.

Thirty-five-year-old naval lieutenant De Long and his party set sail from San Francisco in 1879 on board the steamer *Jeannette*. Like other expeditions of the time, the De Long party included a surgeon, a navigator, a meteorologist, and a naturalist all instructed to collect specimens and record data. Less than two months after its departure from the States, however, the steamer and its team of scientists were caught in packed sea ice. The *Jeannette* spent nearly two years pinned in the ice before finally collapsing under the crushing pressure of the polar pack at about seventy-seven degrees north latitude. When the ship broke up on 11 June 1881, the expedition's officers and crew scrambled onto surrounding ice floes with as many provisions as they could carry. Dividing into three smaller groups, the thirty-three men retreated for the coast of the Lena Delta in search of help. The parties were separated, and only one survived unharmed. In October 1881, the party led by De Long starved to death one by one.[50]

Although the popular press universally lamented the deaths of so many men, several commentators suggested that their lives had been bartered for something equally precious: new knowledge. When the journals of the doomed De Long were published in 1884, for instance, the editor of the two-volume series praised the expedition's contribution to civilization:

> The scientific results obtained were far less than had been aimed at, but were not insignificant. Something was added to the stock of the world's knowledge; a slight gain was made in the solution of the Arctic problem.
>
> Is it said that too high a price in the lives of men was paid for this knowledge? Not by such a cold calculation is human endeavor measured. Sacrifice is nobler than ease . . . and the world is richer by this gift of suffering.[51]

Foreshadowing chemist Edwin Emory Slosson's 1895 remarks, the editor insisted that the relative value of life and knowledge cannot be settled by simple comparative accounting. Even if cold calculations do not determine the real value of Arctic work, the assumption of an underlying system of exchange is plain: the advancement of knowledge carries the price of human suffering.

Of course, middle-class Americans were not unanimously in favor of increasing knowledge through human sacrifice, no matter how willing the participants might be. As in other spheres, the comparative worth of knowledge and human life continued to draw persistent debate. Some of the most biting criticism of suffering for Arctic science came on the heels of the 1881–84 expedition led by Lieutenant Greely, one of the nation's two contributions to the International Polar Year. Of the original twenty-three American men and two Eskimo men sent to the research station in Lady Franklin Bay, only six Americans survived; the others perished slowly from starvation, malnutrition, and exposure. (It is worth noting that the deaths of the expedition's two Eskimo participants, Fred Christiansen and Jens Edward, were not included in the earliest reports of the Greely disaster.)[52] These half-dozen survivors returned to the States with more than two years' worth of systematic records on meteorology, astronomy, magnetism, oceanography, and botany.[53] Critics, however, were not assuaged by the resulting 1,300-page official scientific report. The *Philadelphia Inquirer* condemned the expedition as "monstrous and murderous."[54] President Chester A. Arthur himself declared that "the scientific information secured . . . could not compensate for the loss of human life."[55] The *New York Times* called for an end to the "folly" of such endeavors.[56]

Yet even as critics cursed the barter of bodies for data, their words affirmed a new conception of the relationship between knowledge and sacrifice. Unlike the late Joseph Henry, commentators of the 1880s and 1890s assumed that suffering and polar science were fundamentally linked in that new knowledge must be paid for with the researcher's pain. There remained significant disagreement as to whether the new facts were worth it—that is, whether lives had been wasted (the opinion proffered by most editorials). The terms of the debate, however, had already been set: to advance knowledge of the north, one must pay the price. The quiet, disciplined self-restraint lauded by Henry and his peers had been transformed into a vision of spectacular suffering. The explorer-scientist must be prepared to die for his calling.

The broad reconfiguration of relations among suffering, exploration, and science was carried forward by the valorization of an imperialistic white manhood. As both lay and professional commentators became fascinated with bloody suffering, polar exploration moved into position as the quintessential act of discovery, a metaphor for manly work as a whole. President Theodore Roosevelt, for example, praised the rigors of polar exploration as a corrective to threats of effeminacy and degeneracy. When presenting the Hubbard Medal of the National Geographical Society to Peary in December 1906, Roosevelt declared that it was "a relief" to

> pay signal honor to a man who by his achievements makes it evident that in some of the race, at least, there has been no loss of hardy virtue. . . . We will do well to recollect that the very word virtue, in itself, originally signifies courage and hardihood. When the Roman spoke of virtue he meant that sum of qualities that we characterise as manliness.[57]

Virility, in other words, heralded virtuous morality. White men's spectacular displays of hardihood, epitomized by Roosevelt's famous charge up San Juan Hill, came to demonstrate their honesty and integrity.

The ongoing revision of science and racialized manhood is evident in the expeditions of Cook and Peary in 1908 and 1909. No longer were the quiet, careful, patient observations of gentleman investigators contrasted with the vigor and hardship of exploration. For Cook and Peary, the taking of observations themselves appeared to be grueling tests of physical endurance. Hands bloodied from the wires used to take soundings of ocean depths; piercing headaches from staring too long into the light of the sun reflected in a sextant; feet frozen stiff while hiking to obtain a geological specimen: these and other images of self-punishment fill the narratives of Cook, Peary, and Peary's scientific assistants.[58] Donald MacMillan, one of five white men who joined Peary in the first (but not culminating) leg of his trek to the pole, recalled that they conducted their scientific work "religiously": "lying on our breasts on the sea ice for hours, chipping, chipping ice from the freezing gauge at thirty or forty below zero. . . . [taking] observations every fifteen minutes for six hours and later on every five minutes."[59] MacMillan later described the range of hardships endured in the course of their investigations: "bitter cold, cutting winds, blinding drift, treacherous thin ice, rough ice, pressure ridges, crevasses . . . frost-bitten face, fingers, feet, and starvation."[60] Cook offered a similar litany of hardships: "Privation, cold, hunger, peril of frostbite and of death, solitude, unceasing and sustained exertion."[61]

Yet these early twentieth-century explorers did not merely describe the physical severity of observations and calculations in the Arctic. They insisted on the necessity of such suffering to the advancement of knowledge. Defeating the unknown, they now said, required physical privation. One of Peary's assistants announced that early advances in the north came only through "mistakes which entailed untold suffering and the loss of . . . lives."[62] The freezing of Peary's feet, which led to the amputation of nearly all of his toes, allowed him to accentuate nature's demand for just compensation, as evidenced by his exchange with Matthew Henson, an American who aided Peary on each of his expeditions north. When Henson looked at Peary's twisted and mutilated feet and inquired: "My God, Lieutenant! Why didn't you tell me your feet were frozen?" Peary allegedly replied, "There's no time to pamper sick men on the trail. . . . Besides, a few toes aren't much to give."[63] No longer did the work of science oppose privation; now science itself exacted its own cost; a "few toes" had become "a small price to pay."[64] The "true value" of these expenditures, some explorers stressed, could scarcely be appreciated by "the mind of the average man," who "wants results, tangible results" for his expenditures. In contrast, exploration of the Arctic—"cubic feet of ice thousands of years old, desolate and barren beyond all imagination, [which] will never be used . . . as a productive field for anything to enrich the world in any way"—joined pure science in its utter disutility.[65]

One could here interject that Peary and Cook played up their myriad hardships when addressing popular audiences. Certainly their data—logs of tidal variations, pages of latitudinal calculations, specimens of flora and fauna—bear little explicit reference to sacrifice.[66] Yet it would be a mistake to overemphasize the distinction between science and its representations in this instance. To begin, there is no unmediated access to explorers' raw experience at the pole: their most private documents were intended for public inspection and publication.[67] At every turn, both explorers were aware of the historical significance of their documents and composed their words accordingly. Peary, for example, filled entire notebooks with draft versions of the telegraphs he planned to send back from the Arctic.[68] His assistant MacMillan was horrified when another scientist marred "six hours of painstaking work" by writing a rude remark in indelible pencil in a tidal logbook destined for the Coast and Geodetic Survey in Washington.[69] Even the most mundane details were planned. One awards ceremony became an opportunity for instruction on this point; the National Geographic Society wrote to Peary, "Notify Amundsen & arrange with him when you present the medal to keep quiet, ie *not to move until flash goes off. Flash is planned to go off as you extend the medal & as A. extends his hand to receive it.* Both keep still while this happens & don't stand to [sic] close."[70] The representation of scientific work was a constant preoccupation of the explorers and their backers.

But more important, stressing a distinction between Peary's and Cook's rhetoric of suffering and their "real" scientific work would obscure the fact that the two were fundamentally related in the minds of the numerous experts brought in to determine the veracity of the explorers' claims. Congressmen seeking to resolve the Arctic priority feud, for instance, cast aspersions on Peary's evidence due to the astonishing cleanliness of his diary and record books. A real polar expedition would have been arduous, critics contended; and a reliable journal would testify to this exertion through grease, dirt, and visible wear.[71] Lawyers debating the evidentiary status of Cook's data noted that one might accept his controversial polar account on "the theory that a man having the hardihood to penetrate the Frozen North . . . is too much of a man to claim an honor to which he is not entitled."[72] In the minds of the diverse experts called in to evaluate Cook's and Peary's controversial claims, suffering and credible knowledge were inextricably linked. To prove the attainment of the pole, the explorer's body and records must show the visible effects of the ordeal.

Experts called for such visceral testimony since no other simple method of proof appeared to be possible. Straightforward means of verification, such as observations of the position of the stars or photographs of the stars in the peculiar orbits visible at the tip of the earth's axis, were unavailable. To avoid winter storms and take advantage of the best conditions for traveling over the semifrozen polar ice, both Cook's and Peary's parties traveled in the spring, when the presence of the sun made celestial observations unfeasible. Nor could the explorers deposit cairns, flags, or other objects at the pole for verification by later expeditions. Although both left messages at their respective polar camps, the polar ice shifted continuously with the movement of the sea below, making subsequent verification by inspection impossible.[73] Just as the explorers could leave no stable trace of their presence at the pole, the locale also offered nothing distinctive for them to bring back from its northern axis. Early twentieth-century commentators lamented the absence of polar "Indians" whose abduction from the pole might irrefutably establish an explorer's contact.[74]

Photography proved equally problematic as a means of proving attainment of the pole, despite its status in other realms as a particularly veracious standard of evidence.[75] Although they were picked apart angle by angle in popular and professional writings, the photographs that Cook and Peary brought back from the Arctic hardly solidified their claims for polar attainment. For example, one widely reproduced photograph of Cook's igloo at the "North Pole" also appeared in a book by Cook's one-time associate Rudolph Franke, with no reference to the igloo's situation at the pole. The absence of elongated shadows caused by the low angle of the sun, which must have appeared if the photograph had been taken as Cook claimed, further distorted the photograph's status as a mirror of actual events. The evidentiary status of Peary's photographs was similarly in question.[76]

Written records of polar observations—latitudinal sites, measurements of shadows, oceanic soundings, and so forth—proved to be equally fallible. While several of Cook's and Peary's respective observations and calculations revealed errors when scrutinized by experts, the possibility of mathematical error was not particularly troubling to those who were trying to bestow proper credit on the discoverer of the North Pole. (Peary even suggested that errors further vindicated the reliability of the data since "faked observations will lack the little imperfections which mark genuine observations . . . due to the fallibility of both the observer and his instrument.")[77] The failure of mathematical calculations to prove the truth of either Cook's or Peary's attainments was not simply a matter of error. Rather, the essential trouble with astronomical observations, according to early twentieth-century commentators, was that all of them could easily be forged after return from the pole. Records of shadows, adjusted tables of latitude gained from meridian altitudes of the sun, oceanic soundings, meteorological data—any of these data could be constructed according to readily available nautical almanacs and known principles of mathematics. Peary's colleague MacMillan himself demonstrated the ease of falsification by working out, in full, a series of calculations of his position with respect to the pole. After all necessary corrections, his calculations would seem to place him at a latitude within 152 feet of the pole on 1 May 1928. Yet as he writes in his book, *How Peary Reached the Pole*, on that date he was in fact at Bowdoin Harbor, Labrador, more than 2,000 miles from ninety-degree latitude. As MacMillan concludes from this demonstration, an "astronomical observation for latitude is of the utmost value to the observer, as it proves to him that he has reached a certain spot, *but it is of no value to the world, for it can be easily falsified.*"[78] Thus, problems with records of observations, like photographs, were not thought to reside in observational instruments or techniques themselves. Commentators attempting to sort out the Cook-Peary dispute rarely show concern about the variability of sensitive instruments such as barometers or chronometers but only with the fact that readings from instruments could be falsified at a distance.[79] The underlying problem was proving the integrity of the man behind the machine. Trusting the numbers found in the polar proof, as MacMillan's jest makes clear, required trust in the sincerity of the man who had gathered and presented the data in the first place.

The sincerity of one's fellows was a matter of growing concern at this historical moment. Historian Karen Halttunen argues that new apprehension concerning the deceits of "confidence men and painted women" attended the rapid urbanization, industrialization, and restructuring of middle-class family relations between 1820 and 1870.[80] In the opening years of the twentieth century, this concern with sincerity persisted with a slightly different face. In place of the "confidence man" came the "fraud" or "faker," a person who not only manipulated norms of trust to deceive his fellows but also forged evidence to substantiate his claims and persona. Regular stories of counterfeiting and quackery in popular periodicals and the establishment of boards of experts such as the American Medical Association's Bureau of Investigation point to growing early twentieth-century concerns about deliberate deception and forged evidence.[81]

In the context of wider concern about sincerity, forgery, and proof, suffering assumes novel importance: the explorer's tortured body offered visceral testimony to having "been there." Cook's and Peary's visible decay, the bodily evidence of their grueling ordeal, helped to obviate the need for other forms of proof. Palpable wounds assured myriad readers and observers "that the things had been done and done in the way claimed."[82] After his return from the pole in 1909, Peary immediately hit the popular lecture circuit, highlighting the amputation of his feet, the graying of his hair, and the weathering of his face as evidence of the suffering he had endured en route to the pole. His American assistants published further accounts of the suffering their commander had endured, lending further testimony to the veracity of Peary's achievement. Peary and the five people who accompanied him on the final leg of the journey "were different men" when they returned to the *Roosevelt,* wrote MacMillan. "Their faces, their bodies, their loss of weight, showed plainly the tremendous strain under which they had been. One look was convincing. They had been a long, long way, and had worked hard and had suffered."[83]

As an engineer of Central American canals, a naval commander, and a friend of President Roosevelt, Peary often employed "convincing" suffering to better advantage than did the slim physician Cook.[84] Yet Cook, too, filled his account of the pole with tales of suffering and sacrifice, of "the pain of fasting, all the anguish of weariness."[85] In his first interviews after his return from the north, for instance, he drew attention to the row of teeth he destroyed while gnawing on walrus hide. "The fact is," he stated, "during the last stages of our return journey we were reduced to the very verge of starvation. On some occasions we had to go two and two and a half days without a bite of food. In that period of privation we succeeded in staving off death by famine by eating walrus hide."[86]

Offered as evidence of masticated walrus hide and hence of extreme starvation, broken teeth lent credibility to Cook's claims. The visible evidence of the experience of physical hardship helped to assure the trustworthiness of the explorer and generate assent to his assertions.

Not all suffering, however, offered equal credence. As both Cook and Peary discovered after their return to the States, none of the men in their polar parties was endowed socially with the history of full self-possession necessary to qualify him as an impartial witness. On the final leg of his 1909 attempt at the pole, Peary had been accompanied by five men: his "Negro" assistant, Matthew Henson, and four Eskimos—Seegloo, Ooqueah, and the brothers Ootah and Egingwah. Cook, traveling in 1908, had been joined by two Eskimo men: Etukeshuk and Ahwelah. The fact that neither Cook nor Peary had white companions to corroborate their claims worried even their most enthusiastic and sympathetic commentators. As the headline for one of the first reports of Peary's attainment of the pole noted cautiously, "Peary Tells of Winning Pole: His Only Companions Four Eskimos and Hansen [sic], a Negro."[87] In more critical evaluations of Cook's and Peary's assertions, the testimonies of Henson and the six Eskimo men were simply assumed to be unreliable. Critics brushed off the men's ability to ascertain and record navigational position. Henson, for example, was peppered with insinuations about his literacy and mathematical ability, despite his declared love of Shakespeare and lengthy descriptions of positional calculations.[88] When Henson spoke in public about the expedition to the pole, Peary's white backers instructed him to answer only a limited set of questions from his audiences to quell further speculations about his scientific qualifications.[89] Cook's and Peary's other assistants were also dismissed as competent witnesses. One editorial on the Cook-Peary controversy suggested that Cook's two companions, "untutored Eskimos," were "devoid of the scientific knowledge that would enable them to give intelligent and valuable testimony on such a subject as that under investigation. If they *were* at the Pole, or thought they were, it may have been only because Dr. Cook told them so, and therefore their testimony to the ultimate fact could not add much to his."[90]

Presumptions of illiteracy and lack of navigational skills were not the only aspersions cast on the testimonies of the seven men. Henson's ability "to corroborate or contradict" Peary's statements was dismissed by congressional representatives debating Peary's retirement package because Peary had routinely described Henson as "as loyal and responsive to my will as the fingers of my right hand."[91] Similarly, critics argued that Eskimos were "peculiarly liable to the influence of parties of superior intelligence and craft."[92] The testimony of Etukeshuk and Ahwelah, one prominent legal journal asserted, "must be placed in the same category as the testimony of servants, which, when given in behalf of their masters, is deemed unreliable under another rule of law."[93] "Eskimo evidence is worthless," agreed another expert. "The head of a scientific institute might as well invoke the testimony of his servant in regard to any important experiment."[94] As in the earlier *California v. Hall* decision regarding the reliability of Chinese witnesses, testimonial credibility presupposed full self-possession—the absence of servitude and obligation—and the civic participation it engendered. And as before, these positions were established fractionally and relationally, arranged as points on a continuum rather than as absolute or fixed oppositions: a white male assistant might be historically endowed with more self-possession than a black male assistant, but both held less credibility than the white male chief explorer. Deemed unreliable by their various locations in vestibular selfhood, Henson, Ootah, Egingwah, Seegloo, Ooqueah, Etukeshuk, and Ahwelah were barred from the most decisive sites of the controversy's adjudication.

Just as the words of these seven men were considered unreliable, so, too, their physical trials took on a different significance.[95] The same actions described to inspire white audiences' confidence in the claims of Peary and Cook were treated as evidence of innate racial difference if enacted by Henson or the Eskimos. For example, while white men who endured cold and hunger in the relentless pursuit of Arctic knowledge were lauded for their advancement of knowledge, the Eskimos who took part in the expeditions of Peary and Cook were said to have joined the northern expeditions to "satisf[y] for a time their desire to roam afield, ever persistent in an Eskimo, by nature a nomad."[96] The Eskimos' innate "inquisitiveness," one of the most recurrent tropes of turn-of-the-century polar narratives, is found in the accounts of Cook, Peary, and Peary's American assistants.[97] In a passage that echoes the same litany of hardships (cold, exertion, exposure to the elements) used to extol the white American man's sacrifice for science, we now read of one unnamed Eskimo woman's childlike curiosity about the white men's "strange treasures":

> The . . . Eskimo woman . . . had subjected herself to a temperature of thirty-five degrees below zero, with the liability to be caught in a gale; she had travelled forty miles over a track the roughness of which frequently compelled her to dismount from the sledge and walk; she had carried her child all the way; her sole motive being her curiosity to see the white men, their igloo (hut), and strange treasures.[98]

This woman's endurance of hardship and danger does not demonstrate her sense of higher purpose—a gallant patriotism or a selfless quest for new scientific knowledge. Instead, her travels indicate her people's "childlike" wonder; her physical stamina buttresses the white assumption of native robustness.

Whatever pain Matthew Henson might have endured was similarly erased from both popular and professional accounts of the expedition, despite the fact that Henson was the American party's only fluent speaker of Inuktituk, its best dog driver, and arguably the most crucial member of the team. Henson's blistered feet and frozen face were rarely portrayed as evidence of voluntary and heroic suffering but were offered as further testimony of the natural endurance of nonwhite peoples. Peary captured this perspective in one of his best-selling polar narratives, declaring that, like the Eskimo, "negroes . . . are indefatigable."[99] When Peary and Henson returned to the States from their historic expedition, the symbolic distinction between the white man's "noble suffering" and the black man's "innate endurance" was made manifest in specific material inequities: for instance, Peary drew more than $1,000 per public appearance after their return, while Henson struggled to find a job as a mail clerk earning $2,400 per year.[100]

The explorers' wives were also excluded from the realm of sacrifice for science, albeit in a manner quite unlike the situations of Henson or the Smith Sound Eskimos. The safety and domesticity of the two wives' quiet lives were continuously set in opposition to the perilous work of the manly explorer. Although Marie Fidell Hunt Cook and Josephine Diebitsch Peary had both traveled with their husbands on earlier expeditions—and Peary had even given birth on one expedition—their activities were generally described as wifely endurance. Josephine Peary's own published Arctic narrative features peaceful scenes of cooking, sewing, and waiting for news of her husband interspersed with encounters with polar bears, fatigue, and Eskimo "children of nature."[101] Whether keeping the home fires burning in the States or tending domestic affairs in the far north, the comfortable activities of middle-class white women provided a constitutive counterpart to their husband's spectacular exertions. Arctic narratives, both their own and those of their husbands, tended to contrast their peaceful domestic lives with the exertion and danger of manly scientific work.[102]

Evident, then, is a familiar paradox: the voluntary endurance of physical suffering demonstrated the veracity of Cook's and Peary's claims to having reached the pole, while the voluntary endurance of physical suffering also demonstrated the racial inferiority of Negroes and Eskimos. Suffering was at once one of the benefits of scientific work (for, like imperialist war, it restored men to their healthy, virile selves) and one of the things to be obliterated by scientific work (for *civilization* was partly defined as the superseding of physical toil). Heralded as the means to noble manhood through strenuous activity, science was also promoted as the means to evolving beyond unnecessary physicality.[103]

Indeed, even while Peary's invocation of suffering helped support his claims to have discovered the Pole, he and his allies claimed the *conquest* of suffering as his trademark. In numerous discussions of the "Peary System" of Arctic work, a plan of movement and preparation explicitly likened to Frederick Winslow Taylor's "scientific management" of the industrial workplace, Peary emphasized his triumph over the toil and misery of less rational men. Amounting to a manner of organizing men, dogs, and equipment, the Peary System was said to distinguish the success of this journey from the hundreds of previous painful, failed expeditions to the pole. The lack of scientific planning on expeditions such as Sir John Franklin's had resulted in "untold suffering." The organization and rationality of Peary's plan, in contrast, was said to engineer former hardships out of existence.[104] This depiction of science as a triumph over pain rides alongside its depiction as necessitating pain; again, we see the characteristic ambivalence of these modes of subjection.[105]

MacMillan's best-selling narrative of the 1908–9 journey reproduces this tension. Throughout his book, MacMillan elevates the conventional Arctic tropes of hunger, frostbite, darkness, extreme physical exertion, and emotional exhaustion. "There is no denying the fact that the white men suffered," he wrote:

> As one said, "A hell all right!" All were frostbitten. There were black patches on every face. The rims of our ears were black, where in desperation we had shoved back our hoods to cool our sweating heads and necks. A dull pain across the forehead generally brought us to our senses and caused us to cover up. The tips of our fingers and toes were horny, cracked, and bleeding.[106]

Shortly after this description of physical mutilation, however, MacMillan expressed his hesitation at presenting such spectacles of pain. "It is with reluctance that I include personal suffering in this narrative," he wrote. "It should be endured and nothing said."[107] MacMillan here struggled with conflicting ethics of manliness. Both versions valued the endurance of hardship; the difference was found in communicating that hardship to others.

Warring norms of suffering manhood, not surprisingly, also informed Cook's and Peary's claims to polar discovery. This ambivalence is most apparent in the figure of Cook. Before meeting the press after his return, he trimmed his long, bedraggled hair; scrubbed his dirty and weathered skin to give it a "civilized" sheen; and polished his pemmican-stained teeth. The roughness of polar exploration, valued by Roosevelt as its salient virtue, was tempered by the accoutrements of urbane civilization. Cook's estimation of the importance of a civilized appearance did not go unappreciated. In popular discussions of the polar controversy, Dr. Cook's poses of civilized reserve often increased his appeal, his self-restraint appearing more Christian and charitable than the boasts and threats of the domineering Peary. In newspapers, Cook was often fashioned as the mature and calm negotiator in a civil dispute, as contrasted with a loud and blustering Peary. The front page of the Washington, D.C., *Evening Star,* for instance, carried two articles on the polar controversy on 7 September 1909. An article appearing in column 3, "Peary Ready to Dispute Cook's Claim to be First to Discover North Pole," describes the belligerent Peary's readiness to defend his claim by any means necessary, while an article in column 1, "Cook Wants No Row," presents Cook as gracious and accommodating: "'By going much farther to the east than I did," Cook is quoted as saying, "Commander Peary has cut out of the unknown an enormous space which, of course, will be vastly useful and scientifically interesting.'"[108] Such modesty drew praise. A woman from South Hamilton, Massachusetts, wrote to the explorer to commend his reserve: "Your reticence of speech, your kindly spirit toward a seemingly unreasonable rival, commend themselves to your admiring countrymen. It is a sign of true nobility."[109]

Peary's boastfulness, on the other hand, often drew public rebuke. Even his closest allies recalled that Peary appeared to some to be a "tyrant," a "martinet," or an "autocrat."[110] Peary backers anxiously sought to justify his abrasiveness, explaining that Cook's offensive violation of truth excused Peary's lack of "drawing-room courtesies."[111] In other quarters, the spectacle of voluntary suffering provoked amusement rather than rebuke. One poet, taking an entirely different tack from Elsa Barker's heroic tribute to the links between violence and knowledge, poked fun at the ways in which Arctic suffering was used to shore up the explorer's claims to reliable knowledge. The spoof, titled "Rime of the Modern Mariner," concludes:

"Is's awful and it makes you swear,"
The Mariner began,
"With hunger, hunger everywhere—
And only pemmican."

A tear gleamed in his honest eye:
"Beneath those arctic roofs
I thought for hunger I should die—
And so, I ate my proofs!"[112]

Thus, while sacrifice was coming to be taken as a sign of the white man's moral and scientific integrity, it also provoked guffaws and condemnation. The American public, like Peary and Cook, struggled to accommodate conflicting norms of science, manliness, and suffering.

In short, during the first generation of American exploration in the far north, the grueling physical sacrifice of Arctic work and the cerebral world of natural philosophy were usually opposed. Through the professionalization of American research, the rise of novel international standards of Arctic investigation, and changes in middle-class American values more broadly, science and exploration were reimagined as sharing an essential relationship to heroic suffering. Gaining new knowledge of the Arctic, it was said, required privation and hardship. While the association of science with suffering owed much to the rhetoric of militarism (one Arctic narrative notes that discovery, like war, leaves a field "sown with graves"), the influence worked in both directions.[113] Even as Arctic explorers compared their work to war, warriors began to compare their work to the perils of polar exploration, as Erich Maria Remarque testifies in his memoir of the Great War.[114]

Whereas some voluntary suffering in the Arctic was portrayed as gallant, heroic, and eminently civilized, other, similar actions were said to reveal innate robustness or childlike inquisitiveness. Only the former—the suffering of white men—could be used to establish the integrity of an explorer's scientific evidence. For these men, the experience of absolute north, like the experience of divine presence, was marked directly on the body.[115] Even for them, however, the spectacular demonstration of suffering provoked some ambivalence. While the display of physical hardihood conformed to emerging norms of proper manly behavior, it also conflicted with whites' existing understandings of racial hierarchy (which suggested that the white man must use his civilized mind to *avoid* spectacular physical hardships). As the chapters 5 and 6 reveal, the ambivalent meanings of manhood, civilization, and science troubled investigators outside the Arctic as well, provoking varied responses in different locations.

5

Martyrs

Even more than narratives of polar exploration, accounts of early American roentgenology tend to linger over the ghastly wounds of its participants: ruptured blisters, cancerous limbs, and pus-ridden grafts. These descriptions of decaying flesh are paired with vigorous assertions of experimenters' willingness to suffer: litanies of dismemberment and death glorify rather than trouble X-ray investigators' unflagging devotion to science. In lieu of cool-headed assessments of radiation protection or surgical anesthetics, we find instead an emphasis on a deliberate, even fervent, embrace of further pain.[1] In 1926, for instance, the *New York Times* reported the seventy-second operation on Johns Hopkins University roentgenologist Frederick H. Baetjer, a physician whose years of research with X radiation had already cost him eight fingers and one eye. "Despite the suffering he has undergone in the interest of science," the paper announced, Baetjer planned to "continue his work as long as he lives, fingers or no fingers."[2]

Journalists were not alone in trumpeting investigators' eager return to the source of their wounds. Describing a young physician who continued his research with the ray even as chunks of his fingers, hands, and chest were disintegrating from cancer, a fellow X-ray experimenter averred that the physician's "enthusiasm for his work never faded up to the moment of his end."[3] Even non-fatal X-ray research offered opportunities for fortifying an ethos of energetic suffering. Eminent General Electric engineer William D. Coolidge, for example, was lampooned at one company dinner by "N. Thusiasm," a figure who labored away at new experimental problems "while everyone else slept."[4]

This chapter explores the place of such ardent sacrifice in the first years of roentgenology. Of what significance is this emphasis on enthusiasm (from the Greek *en theos*), with its connotations of spiritual possession, of "god within"?[5] As I will show, American roentgenology gained coherence through a complicated entanglement of knowledge, desire, and visible injury.[6] Much as ritualized religious violence lends immediacy to intangible, supernatural deities or the bloodshed of international war establishes the imagined borders of the state, so, too, the fledgling science of roentgenology acquired definition through the voluntary sacrifice of its adherents.[7] Manifesting the power of an enigmatic new ray and of a novel profession, the willing suffering of roentgenologists invigorated a vision of science as an independent, self-determining force whose advancement depended on loss. At the center of this devotional expenditure was the X-ray investigator, producing the imagined body of science even while suffering in its name.

This sacrificial self, like those already discussed, was shaped by privilege and exclusion. Although alarming numbers of people were killed in the production, distribution, and use of X-ray equipment in the opening years of the twentieth century, only those in certain social positions were listed among the ranks of "martyrs to the X ray." The right to scientific self-sacrifice, however, was never simple or straightforward. The inescapable physicality, and evident utility, of X-ray research positioned early experimenters precariously close to other, socially degraded forms of useful manual labor. Aligned by history with both the practical humanitarian aims of medicine and the demands of low-status wage work, X-ray experimenters were not obvious candidates for an ethos of pure devotion to science. In this context, an emphasis on voluntary sacrifice—on loss beyond exchange value—helped experimenters participate more fully in the more rarified realms of science, endowing their X-ray related injuries with value beyond compensation.

THE ORGANIZATION of a roentgenological profession began shortly after Wilhelm Conrad Roentgen announced his discovery of a "new kind of light" in the closing weeks of 1895. In the earliest years of X-ray experimentation, investigation was unimpeded by regulations on the purchase or maintenance of X-ray equipment. Unlike the later discovery of radium, which was so expensive that only a few individuals and institutions could obtain it, even investigators of fairly modest means might purchase static generators, induction coils, X-ray tubes, and sundry other experimental components. These early investigators, who tended to work alone or in small local groups, flocked to the new field. They brought with them an eclectic array of interests and backgrounds, including surgery, general medicine, physics, photography, and electrical engineering. These disparate groups soon organized, taking a major step forward with the formation of the American Roentgen Ray Society in 1900. The infant specialty was pushed further down the path of professionalization by the publication of the damning 1910 Flexner Report, which stirred reform throughout American medical communities. World War I provided additional stimulus for social and technical expansion, and membership in regional and national roentgenological societies grew accordingly.[8]

Throughout this period, X-ray practice was inescapably manual: investigators gingerly shifted the glass plates on which radiographs appeared, carefully rearranged broken limbs and other objects for better observation, and passed their own hands beneath active rays when testing equipment and modifying dosages. In the twenty-first century, medical and dental X-ray technicians generally leave the room after each adjustment of patient or apparatus before reactivating radiation equipment. In contrast, early twentieth-century investigators often performed each task beneath (or near) a charged X-ray tube. Furthermore, early equipment was testy and fragile, requiring lengthy, repeated exposures for therapeutic or diagnostic effect. Fingers, hands, arms, and faces—the parts of the investigator's body most involved in this meticulous manual labor—were thus particularly vulnerable to radiation damage.[9]

From the beginning, alarm about the dangers of prolonged X-ray exposure accompanied investigators' experimentation. Published reports of hair loss after prolonged exposure appeared within weeks of Roentgen's first public announcement of the new ray. Asked to locate a bullet in the head of a wounded child, Professor John Daniel and Dr. William L. Dudley decided in February 1896 to experiment with the feasibility of skull X rays. Recalling the chain of events later, Daniel reported that Dudley, "with his characteristic devotion to the cause of science," agreed to lend himself to the first trial. Daniel placed his partner's scalp one-half inch away from the ray and exposed it for an hour. Twenty-one days later, all hair had fallen from the area of Dudley's head held closest to the tube's discharge.[10]

Other X-ray investigators soon began to report the drying and inflammation of skin exposed to the ray. Already in August 1896, prominent electrical journals were carrying accounts of more serious effects. An article in the *Electrical Review* titled "Deleterious Effects of X Rays on the Human Body" recounted the experience of Herbert D. Hawks. While a student at Columbia College, Hawks had earned extra money exhibiting an X-ray machine in a New York department store, using the machine to reveal the bones of his jaw to spectators. Demonstrating this "unusually powerful X-ray outfit" for two to three hours at a time over four consecutive days, he was forced to stop work "owing to the physical effects of the X rays upon his body." His hands swelled and took on the appearance of "a very deep sunburn." Two weeks later, the skin fell off his hands, his fingernails stopped growing, and he lost all hair on his face and the sides of his head. The young man's vision was impaired, his eyelashes fell out, and his eyelids swelled. Physicians consulted about the damage compared the symptoms to an instance of "parboiling."[11]

Over the next several months, other electrical investigators, physicists, photographers, and medical doctors who were engaged in X-ray research sent the country's leading journals information about the ray's detrimental effects. Boston tube manufacturer G. A. Frei reported that the skin on his hands and the

hands of a worker, "Mr. K," turned red, hardened, blistered, and fell off after repeated exposure to the ray. In addition, K's eyes burned and ached, and his fingernails felt as "if pounded by a hammer."[12] In September, an investigator from
the University of Minnesota described the "angry sore" that had once been his
forehead, as well as a mouth so cracked, bleeding, and blistered that he could
ingest only tiny bits of liquefied food.[13] In October 1896, the prominent journal
Nature carried a full-page description of the gruesome results of one researcher's prolonged exposure to the ray.[14] By the end of the year, reports of
X-ray burns were front-page news in most prominent electrical, medical, and
scientific periodicals.

Yet whether the ray itself were responsible for the so-called "X-ray burns"
remained a contentious question. Hawks, for instance, dismissed the ray as a
factor in his injuries.[15] He was not alone in looking beyond the ray to explain the
burning effects. Other researchers attributed X-ray injuries to the ozone produced by the apparatus, to a "brush discharge" that occurred when the ray was
held too near the skin, to ultraviolet rays, or to "static changes" that interfered
with the sustenance of the exposed body part. Some suggested that only peculiarly sensitive patients were susceptible to X-ray burns; others pointed to faulty
or weak X-ray tubes or strong tubes operated by untrained personnel.[16]

Perhaps the most strenuous argument over the source of damage took
place between Boston tube manufacturer Frei and electrical investigator Elihu
Thomson.[17] Frei, whose profits from tube manufacturing would undoubtedly
have declined were the X ray shown to be harmful, conducted a number of
experiments in the late 1890s to ascertain the cause of the injuries, concluding
finally that the source of electrical power rather than the rays themselves provoked the damage. After substituting a static machine for induction coils, Frei
decided that the ray produced no "ill-effects" whatsoever.[18] For his part, Thomson initially failed to discern any of the tanning or other effects reported by earlier researchers.[19] But after hearing numerous conflicting reports about the
sources of X-ray burns (including Frei's indictment of the induction coil), he determined it to be "his duty to learn and publish the facts."[20] Like tube manufacturer Frei, Thomson had complex motives: as the developer of an induction coil
for General Electric, he had good reason to argue that the coil was not the source
of the burns.[21] By the end of 1896, Thomson had readied his results for publication. After a short discussion of the matter in an article published in mid-
November, he made a grand entry into the X-ray burn debate one week later,
publishing a lengthy article, "Roentgen Rays Act Strongly on the Tissues," in the
nation's three leading electrical journals, which declared his strong opposition
to the theory that the effect was electrostatic in origin.[22]

To establish his point, Thomson followed Frei in testing the theory on his
own flesh. After exposing a patch of skin on the little finger of his left hand for
half an hour at close range, Thomson tracked the effects. When reporting his re-

sults for readers of *Electrical World, Electrical Review*, and *Electrical Engineer*, he used the detached, depersonalized language that had come to characterize the "culture of no culture" of modern science:

> At present, about seventeen days have elapsed since the exposure, and the finger is still quite sore. . . . Two-thirds of the exposed portion is covered by a large blister which becomes larger each day. The pain and sensitiveness is less after the blistering takes place. The effect has not extended through the finger, but is confined to the back and sides, and is strictly limited to the exposed portion.

Thomson concluded that these painful experiments left him "more than satisfied with the results of my inquiry into the action of the rays."[23]

Like the claims of turn-of-the-century polar explorers, facts about the otherworldly X-ray burns came to rest on the bodily testimony of the "trustworthy investigator."[24] And as before, the disproportionate weight afforded to some bodily wounds in resolving debate over the source of injuries indicates the relational and situated value of the suffering self. Although Frei possessed sufficient testimonial credibility to publish reports of his wounds in nationally distributed journals, the manufacturer's injuries do not appear to have persuaded many of his readers. Nor did the bodily testimony of the student Hawks, whose experience of "parboiling" was reported in the *Electrical Review*. In contrast, the bodily injuries sustained by Thomson, who by 1896 was already an eminent scientist, were mentioned repeatedly in his contemporaries' affirmations of his central findings; the authority of bodily injury enhanced the status of the scientific self, and vice versa.[25] Affirming Thomson's results in *Electrical World*, R. B. Owens expressed his hope that "Prof. Thomson's experience will not prove serious."[26] William H. Greene wrote to Thomson personally from Philadelphia in December 1896: "I was sorry to hear of the sad results of the X rays on your finger, and hope you will make your next experiment on someone else and not yourself."[27] In each instance, an ethereal and ambiguous cause was made visceral and convincing through the fully self-possessed experimenter's injuries. As divine presence was once written on the body of the stigmatized saint, so the effects of the supersensual ray were made palpable in the display of individual wounds.

By late 1896, however, the vitriolic debate between noted electrical and medical experts over the causes of the observed skin effects had waned, partly due to the authority given to Thomson's seeping, growing blister. Moreover, the popular press grew increasingly preoccupied with the ray's burning effect. Reports of the lesions and soreness afflicting X-ray patients and clients appeared within the first year of the ray's use. Popular books already referred to the ray's potential for "destruction" as early as 1897; and by the turn of the century, physicians were compelled to assuage anxious patients before conducting X-ray examinations.[28] By April 1898, popular apprehension about the ray's dangerous "period of incubation" reached such proportions that even the Council of the Röntgen Society, a group unequivocally supportive of the new device, resolved to appoint a committee "to collect information on the subject of the alleged injurious effects of Röntgen rays."[29] Between 1900 and 1902, patients who had been injured by prolonged exposure to the X ray began to sue for damages.[30]

Any remaining doubts that prolonged exposure to X rays might have grave physiological consequences were dispelled by reports of deaths among early ray investigators. Clarence Madison Dally, the first American known to die after prolonged exposure to X rays, succumbed to cancer just as X-ray professional societies were beginning to be established. Born in 1865, Dally was a glassblower employed in Thomas Edison's New Jersey laboratory. He had been investigating rays since the news of Roentgen's discovery first hit the United States in 1896, working primarily on the development of the focus tube and improved fluorescing chemicals. Shortly after Dally began his research, he began to experience the ill-effects of his work. By 1900, he had lost his eyebrows and eyelashes as well as all the hair on the front of his scalp and on his hands and fingers. Because the skin on his hands had grown swollen and painful, he constantly switched hands during his work with the ray. In 1902, after six years of constant pain, Dally underwent the first of a series of grafts designed to relieve the ulceration on his left hand. When the skin failed to graft and examinations revealed carcinoma in the remaining tissue, the hand was amputated above the wrist. The cancer continued to spread, and eventually both arms required removal. Despite the amputations, Dally died from cancer in October 1904.[31] Although Edison began to report X-ray burns in the country's newspapers even before Dally's death, influential critics nonetheless derided the inventor for taking so long to acknowledge the ray's dangers. An editorial in the *Journal of the American Medical Association*, published just before Dally's death, jeered at Edison for "rehashing" facts "learned by physicians through sad experience seven years ago." "He has, as far as can be judged by the newspaper reports, found nothing that has not been found before, and offered no explanation that is of any value."[32] By the summer of 1903, the editors of the journal already believed that the danger of X radiation was old, tired news.

The dangers of X-ray exposure were confirmed by the conspicuous deaths that followed. On 3 August 1905, Elizabeth Fleischmann-Ascheim, known as "the most expert woman radiographer in the world," died from X-ray induced cancer in San Francisco after a series of amputations.[33] One of the first X-ray experts in California and one of the few women in the world known for her work with the ray, Fleischmann-Ascheim gained national renown for her radiographs of U.S. soldiers wounded in the Philippines during the Spanish-American War. On her death, major newspapers published full-page eulogies on "America's Joan of Arc."[34] Fleischmann-Ascheim's death was followed shortly by others around the country, among them Louis Andrew Weigel of Rochester, New York (1854–1906), and William Carl Egelhoff (1872–1907), Wolfram Conrad Fuchs (1865–1908), and Rome Vernon Wagner (1869–1908), all from Chicago. Paired with the range of lay and expert commentary on the ray's destructive effects, these fatalities both augmented awareness that X ray use could lead to severe physiological damage and encouraged early investigators to begin representing themselves as martyrs.[35]

Early X-ray investigators were scarcely the first explorers of the natural world to be memorialized in the overtly religious terminology of martyrdom. Sir David Brewster's 1841 *The Martyrs of Science* and Gaston Tissandier's *Les martyrs de la science* attest to an established tradition of scientific hagiography.[36] In roentgenology, however, inchoate sacrificial ideals found specific expression, a particularly vivid rendering of what chemist Edwin Emory Slosson had hailed as "self-immolation on the altar of science." The new rays—named *X* by Roentgen in recognition of their enigmatic character—confounded simple explanations of cause and effect. Like Henry Adams, prostrated in the Gallery of Machines at the Great Exposition of 1900, early researchers struggled for metaphors adequate to the bewildering new force.[37]

Invisible, active at a distance, and potent beyond any received understanding, the uncanny X ray invited religious comparison. Adopting common Christian tropes, Greene played on these affinities when writing to Thomson in 1896. Referring to the oozing, suppurating blister Thomson developed after testing the ray's action on his hand, Greene queried, "Why don't you help some of the good New England Congregationalists get up a new theory of Hell in which the quivering flesh shall be scorched through and through with these rays which blast and wither but do not consume"?[38] Like divine judgment, the ray seemed invested with eternal, infinite power. Exactly what this power was varied in period texts: ray, force, theos, spirit? In some of the earliest public displays of the X ray, popular observers also remarked on its religious character. At the Edison Fluoroscope Exhibit, some "crossed themselves devoutly after a fearsome glance." One skeptic looked at his own bones and, "with the involuntary ejaculation, 'Oh! My God,' hastened away with the realizing sense of what modern science could do to confirm the predictions of Ancient Scripture as to exposing hidden things."[39]

In the face of this power, some investigators decided to reject its proffered opportunity to suffer. Within the first decade of X-ray experimentation, at least a few investigators began to conduct their research with the aid of protective clothing and devices. Physicians William Rollins and Francis H. Williams, for example, began to shield themselves from the ray even before the first sign of injury. As Williams recalled, "I thought that rays having such power of penetrating matter, as the x-rays had, must have some effect upon the system, and therefore I protected myself."[40] Others, including Edison, abandoned their experiments with X rays altogether and even refused to use medical X rays when ill.[41] Still others, particularly those who began their investigations before the turn of the century, may simply have assumed that their cancers were too far advanced to bother retiring from X-ray work.

Such caution, however, was uncommon in most contemporary accounts of X-ray experimentation. Equally uncommon were accounts of accidental or unwitting suffering. Dr. Emil Grubbé, one of the first American roentgenologists to be seriously injured in the course of X-ray work, later made the rare claim that he was "entirely unconscious" of any hazard during his early experiments. As he wrote in 1949,

> in testing my x-ray tubes, I received massive and uncontrolled doses of x-rays. I was entirely unconscious of any danger. I did not realize that the silent new force which emanated from these vacuum tubes would do irreparable damage to my anatomy. I did not realize what fate had determined for me. I did not realize that I would suffer from the effects of these exposures as long as I lived.[42]

Grubbé characterized his injuries, "unconscious" though they were, as a "sacrifice . . . to medicine and the sick."[43] His suffering here appears harrowing but both inevitable and productive. As a result of the scientist's unintentionally painful accumulation of knowledge, society passes from ignorance to knowledge, from suffering to salvation. "Nature has always been merciless in her demands of the workers in the vineyards of science," Grubbé summarized. "Every step in progress costs something; and, ultimately, the price of progress is usually very high."[44]

Grubbé's unusual account of unwitting sacrifice represents one version of the era's underlying presuppositions of contract. In his portrayal, there exists a regulated and rational system of remuneration (albeit one veiled from the individual researcher's own comprehension), in which the flesh of X-ray experimenters is payment for knowledge received. In this sense, Grubbé's narrative provides a compensatory vision of exchange based on an implicit contract with nature. As Emerson would have insisted, "in nature nothing can be given, all things are sold."[45]

This narrative of reciprocal exchange, relatively uncommon in early twentieth-century discussions of roentgenology, has since become a favored device of more recent historians. For example:

> The harmful effects of X-rays . . . work insidiously with the passage of time, hour by hour, day by day. The organism records even the tiniest absorption with the precision of an electronic machine; like a faultless book-keeper, it adds each day's sum to the preceding one, and relentlessly totals them up. The bill, it is true, can be renewed with long peri-

ods of rest, if they are taken in good time, but once certain limits are passed, payment at a cruel price will surely be exacted.[46]

In such accounts, roentgenologists' losses are still understood as part of an exchange relationship, although outside the realm of the consensual, voluntary contract. Surrendered limbs and lives are ultimately balanced by the multiple benefits of new knowledge about nature's forces: a "small price to pay." As roentgenologist Lewis Gregory Cole said in 1917 when describing the radiation-induced death of a young colleague, "rarely does the sacrifice avail so much"; "his interest in the relief of the suffering of others led him to forget his own danger."[47]

When we examine the writings of early investigators more closely, however, narratives of cautious self-protection and those of reciprocal, functional exchange (danger to oneself exchanged for the relief of others) both seem peripheral. When other roentgenologists echoed Grubbé in pointing to the cost of progress, they departed sharply from his claims of ignorance and unconsciousness, stressing instead the importance of intentional, deliberate suffering. Most writing on roentgenology in the first decades of the twentieth century emphasized that real dedication to science—true martyrdom—was demonstrated by volition, an active, deliberate decision to pursue investigations known to be harmful to the self. In this voluntarism, roentgenologists were explicitly likened to other suffering scientists. "Like the explorers of unknown countries who suffer privations and the pangs of hunger and thirst," wrote roentgenologist P. M. Hickey when reflecting on the early years of the field, "so our predecessors in American roentgenology frequently and willingly brought upon themselves subsequent sufferings."[48] These accounts tend to move beyond visions of strict reciprocity, instead describing sacrifice as enthusiastic effort given without calculated estimation of returned value. The suffering itself appears paramount.

In one such instance, Philadelphia physician Charles Lester Leonard (1861–1913) underwent one amputation after another (first a finger, then the hand and forearm, then the upper arm at the shoulder) before succumbing to cancer in 1913 at the age of fifty-one.[49] His pain, however, seems merely to have intensified his fascination with the ray. Even after publishing warnings on the dangers befalling the X-ray specialist, he continued his radiological investigation of the urinary tract.[50] The ten years of his most prolific research coincided with the ten years of his "steadily increasing physical distress."[51] Another investigator, surgeon Stephen Clifton Glidden (1870–1917) continued his work with the X ray following the amputation of several fingers in 1907. After further amputations made surgical practice impossible, he began other investigations with the ray. Even the amputation of his arm at the shoulder did not dissuade him from his research, and in 1916 he began additional experiments with radium. Forced to cease practice later in the year due to ill health, Glidden died on 20 February 1917.[52]

Investigator Mihran Krikor Kassabian (1870–1910) continued his X-ray work for ten years after publishing a report on its hazards, "X-Ray as an Irritant," in a prominent journal of roentgenology.[53] In 1903, even as the injuries to his hands grew more severe, Kassabian accepted a position in the Roentgen-Ray Laboratory at Philadelphia Hospital.[54] The itching, toughening, and blistering of his skin continued, yet Kassabian persisted in his diagnostic and therapeutic practice. By 1908, after a decade of relentless exposure to X radiation, he was in terrible condition. The fourth finger of the right hand was completely covered with horny tissue, and the left hand had an open sore running across the area between the middle and fourth fingers. In April, the fingers were amputated to stop the spread of X-ray induced cancer. During the following year cancer appeared in Kassabian's armpit, and his axillary glands were surgically removed. When this wound failed to heal, surgeons cut further tissue from his chest. Throughout the period of these surgical excisions, Kassabian continued his X-ray practice, ceasing only when too weak to work. In 1910, as his important textbook on X rays was appearing in its second edition, he died of cancer.[55] Glidden and Kassabian were followed by others. By 1911, more than fifty such cases of X-ray–induced cancer had been reported; and by 1949, at least sixty-five Americans had perished as a result of their work with the rays.[56]

Of course, such statistics are hardly remarkable when compared with other occupational hazards of the early twentieth century. To take just one example, between 1890 and 1917, 72,000 railroad employees were killed on the tracks.[57] Yet both contemporary commentators and later historians endowed these few dozen X-ray deaths with significance beyond mere numbers. They became, in both the period press and the historical record, "martyrs to science" distinguished by their willingness to suffer and die for their cause. Hickey and others who highlighted the deliberate nature of roentgenologists' self-injuries understood the importance of voluntarism to sacrificial ethics better than Grubbé did. Again, the willingness to suffer is the principal feature of these narratives: the scientist must *choose* his distress, lest his actions lose their greater meaning.

Perhaps the most spectacular accounts of suffering in the annals of early roentgenology surround the work of Walter J. Dodd, one of the most important American figures in the field.[58] "The name of Walter Dodd," wrote one commentator, "is as infrangibly interlocked with the course of early American radiology as is that of Rontgen [sic] himself an integral part of the scientific annals of Germany."[59] Already by November 1896, less than a year after Roentgen's discovery of the X ray in his Würzburg laboratory, Dodd was tormented by severe dermatitis. Within five months, the pain was "beyond description," and his face and hands appeared as though scalded. Dodd's employer, Massachusetts General

Hospital, placed his name on a list of those with grave injuries. When the pain in his hands kept him awake at night, Dodd paced the floor of the hospital pharmacy, hands raised over his head.[60] In July 1897, he received the first of a series of skin grafts performed by his friend, Dr. Charles Allen Porter of Boston (soon to become the nation's leading expert in the treatment of chronic X-ray injuries). The grafts were unsuccessful, however; and by 1902, cancer had spread through his fingers. Thus began the effort to salvage Dodd's "useful hands":

> There were fifty operations under ether, which lasted from an hour and a half to three hours. The capable fingers were taken away bit by bit. Rather than yield a fragment of a joint, Dodd would endure the agony of keeping it for months after it should have been removed . . . [sometimes] Dodd went to the operating-table without knowing how much of his hands would be left when he awoke from the ether.[61]

As Dodd and his surgeon evaluated how much to excise from each lesion-ridden hand, they based their judgments on the finger's relative usefulness. Despite the agonizing pain caused by the exposure of raw nerve endings in a cancerous ulcer, Dodd put off the amputation of his little finger for months, fearing that its loss would leave his thumb nothing to press against—rendering it impossible for him to grasp the controls of the X-ray equipment.[62]

At first glance, this emphasis on utility would seem to evoke a conception of the white male body as a machine, a collection of interchangeable parts subject to dismantling and reconstruction. This utilitarian, production-oriented vision was widespread in early twentieth-century America, as recounted in a number of recent histories of the period.[63] Yet in crucial aspects, Dodd was far from this instrumental vision of manhood, neither so stoic nor so orderly. Dodd (at least the "Dodd" recalled by biographers and colleagues) *embraced* his suffering. Unlike an impassive machine, he was said to have "loved difficulty" and to have sought it out for its own sake.[64] Although painkillers would easily have made the performance of his work less excruciating (and perhaps more efficient), Dodd refused narcotics, even his physician's ordered injections of morphine.[65] He typically took only aspirin for his wounds—wounds which his surgeon insisted must have caused more constant, intense, physical suffering than any others he had seen.[66] Upon awakening from surgical ether after yet another amputation, Dodd always returned immediately to his radiological work.[67] "Everybody warned him when the danger became evident," recalled one colleague, "but he would not give up his work and was always eager for the next case."[68] "No man," wrote his biographer, "was ever more enthusiastically absorbed in his task."[69] Even after Dodd's face had grown parchment-like "from the fibrillary contractions of old scars," wrote another friend, "[it] was never too inelastic to break into a smile."[70]

Certainly these descriptions of Dodd's happy suffering must be understood as part of the era's funereal genre, lush praise given in memory of a beloved colleague. Yet in their exaggeration, the words illuminate important norms of manliness, science, and sacrifice present in the early twentieth century. Hardly machine-like or impassive, Dodd was lauded for his anti-utilitarian embrace of pain. Although in other circumstances such an embrace might indicate barbarity (for example, the "ancestral savages" of Lewis Henry Morgan) or pathology (such as the "masochism" of Richard von Krafft-Ebing), in this case a sanctioned affiliation with science circumvented such connotations. Whether actions were to be classified as insane or admirable, degenerate or enlightened was situational, depending on the ever-changing relation between self and science.

Rather than appearing atavistic or perverse in the eyes of colleagues (or subsequent historians), Dodd's voluntary injury helped augment the science for which he suffered: an embrace of pain helped to establish the profession. Dodd's unswerving love for the agent of his destruction appears to have made him an exemplary roentgenologist.[71] As one X-ray expert recalled, "Dodd won the respect and affection of everyone on the M.G.H. staff by his careful, painstaking work and by his ever-willing self-sacrifice."[72] Another noted that he "well knew that he was exposing himself, without adequate protection, to a peril that for him was even greater than shell fire in the trenches. Yet . . . his spirit of self-sacrifice was unhesitating."[73] Dodd's enthusiasm for pain was treated reverentially by his fellows, who described him as a "'roentgen' Saint."[74] According to some recollections, his cheerfulness not only won him respect and affection but also increased his scientific abilities. For example, when Dodd counseled patients reluctant to undergo roentgen treatment or examination, he offered stories from his own experience. "Touched and strengthened by the visible evidence of his own suffering," patients would acquiesce to the recommended diagnostic test or treatment.[75] (One might imagine that patients would *refuse* treatment after hearing stories of the physician's own anguish, but Dodd instead appears to have inspired his patients.) His biographer described an instance in which a "sturdy" young man petulantly refused a requested radiological examination. One visit to the scarred and amputated Dodd ended all "boyish grumbling." The boys who passed through the X-ray room "met a man, and they would have stood on their heads for him."[76]

This enthusiastic rush toward death and dismemberment was a formative aspect of the eclectic field known as roentgenology; the fledgling community identified itself through a shared "spirit of sacrifice."[77] Nowhere is this identification more evident than in the official martyrology crafted by Dr. Percy Brown, one-time president of the American Roentgen Ray Society. Described by a colleague as one of the "great living suffering martyrs from this cause," Brown carefully demarcated the field according to the embrace of sacrifice.[78] He insisted that X-ray investigators understood the prospect of danger within ninety days of Roentgen's first announcement. Armed with "ample warning" and a full "chance to retire," many individuals abandoned their work with the ray. Those who did, according to Brown, included (1) profiteers who hoped to "exploit the x-rays" in popular exhibits, (2) physicists who understood that they had "paved the way for the practical usefulness of the roentgen-rays" and hence turned to other investigations, and (3) medical practitioners who recognized the need to "relinquish" the ray to further development in "special hands."[79] In Brown's account, those who remained in the field despite published warnings were the true founders of the science, "avowed and determined special workers, pioneers in the cause of their Work and in many instances ultimate martyrs to it."[80] Brown thus aligns the boundaries of roentgenology with a will to sacrifice: "special workers" willing to endure pain became "scientists." Those who left, by definition, did not.[81]

The standard of martyrdom adopted by Brown and his colleagues served several functions for the emerging professional community. Most obviously, the promotion of voluntary suffering helped alleviate concern that practitioners would leave the field. Retrieved as noble and heroic, the deaths of the field's founders preserved and increased the status of the profession as a worthwhile calling and promised to attract new members. Evoking the religious language of martyrdom also shielded existing practitioners from some popular criticism about unnecessary waste of life. "By making the victims of X-ray induced carcinoma martyrs to science," Daniel Serwer summarizes, "the medical radiological community could hope to justify the loss of life and also thereby off-set . . . negative public reaction."[82] In the context of a larger valuation of certain kinds of voluntary suffering, turning destruction into glorious sacrifice helped to protect a newborn profession.

Yet it would be an unfortunate misreading to say that professionalizing investigators simply employed martyrdom as tactical rhetoric. Indeed, sacrifice acquired material form in the tattered flesh of X-ray investigators. As stigmata render palpable the ineffable presence of divinity, so, too, scarred and limbless roentgenologists came to embody the abstract cause of science. By the late 1910s, roentgenologists had come to be known by their mutilated limbs.[83] At one 1920 professional gathering, historian Bettyann Holtzmann Kevles reports, so many attendees were missing at least one hand that, when the chicken dinner was served, no one could cut the meat.[84] Participants in this and other macabre scenes forged a community not only through the inflated language of martyrdom but also through the lived experience of dismemberment and pain. Injury became the basis for scientific embodiment.[85] As wizened relics conveyed the prospect of eternal life to the faithful or the bodies of wounded soldiers communicated the significance of the nation, so were the disintegrating bodies of individual x-ray investigators reclaimed as pieces of the boundless unfolding of a "new department of science."[86]

The relationship between roentgenology and injury was made manifest not only in the missing fingers and hands of the living but also in monuments to the

dead. Like war memorials used to consolidate national identity, the construction of public monuments solidified the unity of this fledgling science. A striking example was an official tribute to the "martyrs of radiology" erected in the
garden of Hamburg's Saint Georg Hospital in the spring of 1936.[87] The monument's single vertical stone might be read as a symbolic recovery of the individual investigator's amputated finger, revivified as part of the larger, undying
body of science. The names of individual dead are inscribed on its sides in al-
phabetical order, de-emphasizing differences in rank and status and enhancing
the investigators' shared scientificity. As with so many war memorials, the monument emphasizes collective contribution to the higher goal—in this case, a
shared contribution to the new science of roentgenology.[88]

Such memorials to X-ray sacrifice systematically ignored all but full-time
X-ray investigators. Also succumbing to ray injuries were equipment salespeople, who, like shoe-fitting clerks, repeatedly demonstrated the efficacy of X-ray
merchandise for potential customers. Boston salesman Frank Howard Swett, for
instance, died in 1929 from complications resulting from X-ray exposure.[89] But
neither these merchants nor others exposed to X radiation garnered the high
title of "martyr." Although interested researchers frequently experimented not
only on laboratory animals but also on their wives, patients, and other human
assistants, none of those subsequent injuries or deaths were recorded in the annals of scientific self-sacrifice.[90] With the important exception of one woman,
Elizabeth Fleischmann-Ascheim, the only "martyrs and pioneers of radiological
science" to feature in these descriptions were men employed as full-time X-ray
investigators.[91] (References to Marie Curie's exceptional sacrifices peppered
American accounts after her death in 1934.)[92] By dubbing Fleischmann-Ascheim
"America's Joan of Arc," roentgenologists and the wider popular press preserved
the status of the field as a sacrificial scene without troubling its overarching
masculinization: roentgenologists remained manly "soldiers for science," with
the lone exception of this beautiful Jeanne d'Arc. (Fleischmann-Ascheim's Judaism appears not to have hindered those who identified her with the young
saint.)[93] Although technicians were occasionally included in the rosters of the
dead—particularly the first American martyr, Charles Madison Dally—those said
to possess a true "sacrificial fidelity" typically were men with advanced degrees
in medicine or science.[94]

In similar fashion, sacrifice was narrowed to the perils of radiation, ignoring all other dangers of X-ray work. Despite equal deadliness, none of the other
hazards of roentgenological practice (such as the flammability of the film; the
fragility of the large, chemically treated glass plates; and the potential for electrocution from high-voltage equipment) produced a martyr to science in early
twentieth-century accounts.[95] A 1929 fire in a single X-ray storage area killed at
least 124 people, a figure comparable to the entire number of investigators who
succumbed to X radiation in the United States.[96] No monuments were raised for

these dead, no glowing hagiographies composed, though countless industrial workers were surely injured by electrocution, broken glass, the movement of heavy equipment, toxic chemicals, and industrial fires. To preserve the special distinction of the field of roentgenology, only those workers succumbing to irradiation counted as willing sufferers.

Roentgenology was thus characterized as a pursuit for which certain individuals endured bitter pain and disfigurement. *Roentgenologists*, by implication, were those rare individuals willing to take on the dangers of radiation. Just as the reverence of believers maintains divine power or the devotion of individual citizens preserve the boundaries of the nation, so the enthusiasm of the suffering experimenter sustained a new science. Far from simply supplanting religion, then, late nineteenth-century physical research adapted age-old customs of bodily mortification. The cause of roentgenology was consecrated with each finger removed.

Certainly, sacrifice was not the only norm of scientific advancement available at the time: recall the fortuitous scientific "accident" used to explain Antoine-Henri Becquerel's 1896 discovery of natural radioactivity discussed in the introduction. Furthermore, one can easily envision a professional community that took the idea of willful suffering as a problem to be understood rather than a value to be reproduced; consider, for instance, psychoanalysts' studies of masochism, proliferating even as Dodd enthusiastically returned to the source of injury. Despite other possible formulations, this community of practice twined pain and the pursuit of new knowledge, building a scientific field on an ethic of voluntary suffering. Forfeiture of the self acted as the mortar of the collective. Like America after the Civil War, the legitimacy of this imagined body—a sense of its definition, autonomy, and purpose—was secured partly through destruction: through the considered squander of life and limb.

6

Barbarians

In the closing paragraphs of Sinclair Lewis's 1925 novel *Arrowsmith*, Martin Arrowsmith and his male companion, Terry Wickett, loll in an ungainly boat in the middle of a Vermont lake. Wickett, a chemist, and Arrowsmith, a bacteriologist, have recently rejected the stifling luxury of the McGurk Institute of Biology in New York City in favor of a homemade laboratory and a rickety shanty far off in the woods. Removed from the constraints of urban institutions and urbane colleagues, they sit beneath the evening sky and discuss the progress of their studies of quinine derivatives. "'I feel as if I were really beginning to work now,'" Arrowsmith says to both himself and Wickett. He continues with the sentences that close the novel: "This new quinine stuff may prove pretty good. We'll plug along on it for two or three years, and maybe we'll get something permanent—and probably we'll fail!"[1]

Arrowsmith's prediction of his own failure raised eyebrows among Lewis's readers. When the author first presented the completed manuscript to his friend and collaborator, Paul de Kruif, de Kruif balked at the concluding sentence. Himself a renegade bacteriologist who became a writer after leaving his position at the Rockefeller Institute, de Kruif had worked with Lewis on the novel since its inception, providing crucial concrete details about the practice of biological research. Late in the summer of 1923, when Lewis invited de Kruif to comment on a draft of the novel, de Kruif found the manuscript both thrilling and true to life. "You've done it, my boy. It's great," he recalled telling Lewis. "I'd change only one word in the last sentence where Martin and his friend have gone to the bush to make science on their own." "How'd you change it?" asked the author, whom de Kruif referred to as "Red."

"Those last words—'and probably we'll fail'—"
"What would *you* say?"

"I'd suggest *possibly*, not probably, we'll fail," I said mildly.

"That shows you've missed the whole meaning of *Arrowsmith*," Red said, shaking his head sadly.[2]

In this chapter, I argue that what de Kruif missed in Lewis's representation of modern experimental science was the novelist's vision of science as an essentially unproductive venture. Failure is not epiphenomenal to research in *Arrowsmith;* it is central to the very definition of the true scientist's endeavors. Fundamentally nonutilitarian, Lewis's science confounds the desire for consummation, production, and achievement that de Kruif took for granted.

Like the purists' search for an ever-receding truth, the explorers' quest for the metaphysical pole, or the martyrs' enthusiastic embrace of suffering, *Arrowsmith* valorizes the scientist's uncompensated exchange. Throughout the novel, the protagonist's labors are unremunerated. He belligerently resists all useful scientific work, forgoing both professional publication and therapeutic applications as he takes on obscure bacteriological problems. As if to drive home the significance of the scientist's lack of productivity, Lewis depicts Arrowsmith's social and sexual relations as equally fruitless when gauged by the conventions of the day. Arrowsmith ignores every opportunity to accrue wealth and social power, rebuffing expectations of middle-class mobility. His first marriage proves barren. (His wife's only pregnancy ends with a traumatic stillbirth.)[3] His second marriage produces a son, yet Martin abandons the child to focus on his scientific work.

Taking the novel as an endpoint, this chapter explores Lewis's redemption of such unproductive expenditure. In dwelling on this aspect of the story, I depart from previous treatments of the well-discussed novel. Both literary critics and historians of science have explored *Arrowsmith*'s emphasis on asceticism and science. Yet they have not elucidated the connections between those intriguing closing lines and the novel's religious and scientific themes. For example, in his superb discussion of the novel, historian of science Charles E. Rosenberg mentions the exchange between de Kruif and Lewis but does not explore its broader implications. Referring to the conversation only in a footnote, Rosenberg suggests that Arrowsmith's denouncement of success can be traced to Lewis's feelings of revulsion about the price of his own fame and fortune: "no one had experienced more acutely than he the bitterness of American success."[4] Lewis's biographer, the literary critic Mark Schorer, dismisses the ending as "a little fantastic, and quite unpersuasive."[5] I, however, want to stress the importance of that ending, showing that the protagonist's avowed rejection of intellectual and social productivity appears not perverse but virtuous. While the novel's treatment of sacrifice does not necessarily indicate period consensus (de Kruif's confusion over his friend's choice of endings seems to suggest that Lewis's portrayal of science was not universally shared), it does illuminate larger trends in the history of voluntary suffering.

As in previous instances of self-sacrifice, the novel's evocation of uncompensated expenditure presupposes a state of privilege: only the fully self-possessed man might become the deliberate barbarian Lewis heroizes. Through the characters of Martin Arrowsmith's wives, his institutionally dependent colleagues, and the racialized inhabitants of a fictive West Indian island, the novel establishes the familiar oppositions and exclusions defining the self-possessed subject: contrasts between masculine, sacrificial science and unscientific, feminine domesticity, and between colonial subjection and liberal self-determination. At the same time, the novel twists these patterns in surprising and complex ways, presenting Arrowsmith's renunciation of worldly practicality as a position of contradiction, a life at once squalid and transcendent, beneath civilization yet superior to it. Quite unlike the scientists discussed in previous chapters, the fictional Arrowsmith renounces civility itself. In this portrayal, the free and virtuous scientist must be barbaric. Far from establishing his increasing refinement, Arrowsmith's expenditures highlight his laudable slide toward degeneracy. In this sense, the book marks the culmination of nineteenth-century norms of sacrifice.

AS THE TITLE suggests, *Arrowsmith* focuses on the experience of the individual scientist. It is above all a "biography," a coming-of-age story that describes a young man "who regarded himself as a seeker after truth . . . who stumbled and slid back all his life" (45). He is exceptional only because of his intense interest in scientific research. Born in 1883 in the small midwestern town of Elk Mills, Winnemac (a state said to be bounded by Michigan, Ohio, Illinois, and Indiana), Martin is apprenticed at age fourteen to the kindly town physician, Doc Vickerson, who urges him to attend college before going on to medical school. (Martin's parents quietly disappear by page II, leaving enough money to pay for Martin's college and medical school tuition.) While studying at the University of Winnemac, Martin is inducted into the alluring world of scientific research through eminent bacteriologist Max Gottlieb.

More than any other character in the novel, Gottlieb reveals Paul de Kruif's collaboration, combining attributes of German-American physiologist Jacques Loeb, with whom de Kruif worked at the Rockefeller Institute, and bacteriologist Frederick Novy, with whom de Kruif studied at the University of Michigan.[6] Enlivened by technical and imaginative details provided by de Kruif, Gottlieb is one of the novel's key figures; Lewis even considered naming the book *In the Shadow of Max Gottlieb*.[7] An embodiment of the ideal of research, Gottlieb serves as Arrowsmith's scientific superego: when Martin wavers from his quest for truth, Gottlieb's judgmental voice echoes in his head. Gottlieb becomes an "obsession" for Arrowsmith—the ideal incarnation of "the barbarian, the ascetic, the contemptuous acolyte of science" (369).

Despite Gottlieb's profound influence, Arrowsmith veers from the narrow path of bacteriological research. He marries a young nurse, Leora Tozer. At the urging of the dean of the Winnemac medical school, T.J.H. Silva (whom the medical students call "Dad"), Arrowsmith completes his degree and moves with his new wife to North Dakota. After a short, unsatisfying stint as a country doctor, Martin takes a series of other nonresearch positions: in Nautilus, Iowa, he joins a blustering booster, Almus Pickerbaugh, in the public health department; in Chicago, he joins a lucrative medical practice, the Rouncefield Clinic, as a private physician. Gottlieb then reenters the narrative, inviting Arrowsmith to abandon public health propaganda and clinic greed for a life of research at New York's gleaming McGurk Institute.

While at the McGurk, an institution shaped by de Kruif's recollections of the Rockefeller, Arrowsmith contends with unpleasant professional politics and mobilization for the Great War. In spite of these distractions, he discovers a bacteriophage that inhibits the growth of certain strains of staphylococcus. Encouraged by his peers, he travels to the fictional West Indian island of St. Hubert to test the efficacy of his phage on the island's plague-stricken populace. At first Arrowsmith seeks to conduct a controlled scientific experiment, in which half the exposed population would be left untreated. While on the island, however, both Martin's colleague, the buoyant Gustaf Sondelius, and his wife, Leora, also succumb to the plague. Stricken by these deaths, Arrowsmith abandons his experiment and administers the treatment to all who ask for it.

After his return to the States, Arrowsmith recovers from his grief and marries a wealthy widow, Joyce Lanyon. They have a child, and Arrowsmith flits briefly through the highest circles of New York society. Yet in the final pages of the novel, Arrowsmith abandons wife, child, and social position and flees to Terry Wickett's makeshift laboratory, where he at last dedicates himself to a life of inquiry for inquiry's sake.

The plot conforms to a typical Lewis pattern: an idealistic protagonist glimpses values beyond the confining spheres of his or her immediate environment and struggles to enact them.[8] Other famous Lewis characters, from *Main Street*'s Carol Kennicott to *Dodsworth*'s eponymous lead, also struggle to maintain their personal integrity in a world of corrupt and corrupting values. In *Arrowsmith*, Lewis perfects the narrative of individual struggle by including a heroic retreat to nature, a theme prevalent in American literature since James Fenimore Cooper's Leatherstocking novels. As scholars have been quick to point out, by stressing Arrowsmith's purifying retreat to nature, Lewis revisits one of the dominant motifs of American fiction.[9]

Yet the novel is unique in American literature of the nineteenth and early twentieth centuries in situating those heroic struggles within the realm of science.[10] Certainly scientific protagonists had appeared previously in American fiction: consider Nathaniel Hawthorne's Rappacini (1844) and Aylmer (1845), Edward Bellamy's Dr. Heidenhoff (1880), and the mad tinkerer of Jack London's "A Thousand Deaths" (1899).[11] *Arrowsmith*, however, is the first significant

American novel to feature a research scientist as its central character. As such, the prize-winning novel not only heralded a new literary investment in science but also crystallized ethics of sacrifice for science that had been developing over the previous four decades.

Among pure scientists of the 1880s, polar explorers of the 1890s and 1900s, and roentgenologists of the 1900s and 1910s, science was generally represented as socially and materially unrewarding. Proponents of pure science proudly trumpeted their disgust for lucrative applications; MacMillan, Peary, and Cook emphasized the addictive, obsessive aspect of their quest for the pole; early X-ray experimenters heralded their colleagues' fatal pursuit of new knowledge. The supposed virtues of these pursuits were at once achieved and confirmed through suffering: investigators' voluntary endurance of physical, mental, economic, or social hardship attested to their affective and moral character: pure, trustworthy, noble, and so on.

These starkly different examples, however, shared an emphasis on the civility of their unproductive work. As we have seen, a dedication to "science for science's sake" was said to demonstrate an advanced level of cultural development. In Henry Rowland's words, only an elite few with the "ability" and "taste for higher pursuits" were willing to sacrifice themselves for greater knowledge.[12] According to Peary, his advanced system of exploration would elevate the field; by circumventing unnecessary, unwilling suffering, he revealed the progress enabled by science. The X-ray martyrs similarly presented one another as embodying the highest cultivation of manly virtue. In other words, voluntary suffering not only required but generated refinement. Self-sacrifice both elevated the scientist and demonstrated the lofty moral and intellectual position that a few privileged bodies had already been allotted.

Lewis's *Arrowsmith* at once extends and reverses this pattern. On the one hand, Arrowsmith shares a disdain for material concerns and similarly vows to transcend such interests through a devotion to truth for truth's sake. On the other hand, Lewis turns the civilizing mission of science on its head. Although, like Henry Rowland, he presupposes progressive cultural development through the familiar hierarchical categories of savagery, barbarity, and civilization, Lewis's protagonist heads down the great chain of being in the opposite direction. In stark contrast to Rowland's 1883 declaration that the pursuit of knowledge represents the pinnacle of gentlemanly refinement, western achievement, and moneyed good taste, Arrowsmith's quest for truth appears not civilized but feral. In Lewis's depiction, science requires this barbarity. The scientist's transcendence emerges, paradoxically, through his descent from civilization. The novel's supporting characters also reverse conventional evolutionist rhetoric. Dr. Marchand, for instance, the Negro physician Arrowsmith encounters on St. Hubert, epitomizes urbane, genteel civilization; in Lewis's hands, however, civility becomes the thing that courageous men avoid. Civilization implies a range of attitudes and behaviors in the narrative: table manners, charity, polite language, compassion, concerns about personal cleanliness, and the acquisition of private property. Science entails a renunciation of them all.

Descriptions of Arrowsmith's scientific incivility permeate the novel, which Lewis tried to title *Barbarian*.[13] While serving as acting director of the Nautilus Department of Public Health, for instance, Martin sloughs off learned niceties while investigating the production of hemolysin in sheep's blood. He growls at his wife and at his assistants. He fumes, rages, and sweats (248). He curses, forces his wife out of bed to help him prepare media, and is violent to his stenographer (248–49). Eventually, he is removed from his position as acting director. It remains unclear, however, whether Martin's barbarism is acquired or inborn. In some instances, the scientist-as-barbarian appears to reveal degeneration (that is, he loses his civility through Lamarckian devolution); in others, he demonstrates atavisim (he has somatically reverted to barbarity even before birth). In either case, barbaric he remains.[14]

Nevertheless, Arrowsmith is tempted repeatedly by the delicacies of civilization. With his move to Chicago, he temporarily regains a veneer of gentility. He and Leora enter "the world of book-shops and print-shops and theaters and concerts. They read novels and history and travel." They chat with "journalists, engineers, bankers, merchants" (260). Yet the moment Arrowsmith receives another opportunity to pursue research, his barbarism returns. When he discovers the staph-eating phage while at the McGurk, he forgets "Leora, war, night, weariness, success, everything" (296). As he delves deeper into his experiment, he begins to steal laboratory materials from Gottlieb and cigarettes from technicians (297). He stops bathing. He babbles (300). He stomps around the institute, "pillaging" for food (298). He enters his home only rarely and briefly, just long enough to gobble food "like a savage" (300).

As Arrowsmith's research continues, he spirals still further from norms of civil behavior. He jettisons not only the superficialities of table manners and shaving but even his own mental stability.[15] He becomes neurasthenic and obsessive. He spells words backward. He begins a checklist of his myriad phobias: agoraphobia, claustrophobia, pyrophobia, siderodromophobia (303–4). When at last anthropophobia sets in, Arrowsmith retreats from his work in an effort to regain self-command (304). Only with the timely intervention of the McGurk administrators—who demand that Arrowsmith cease experimenting in order to publish—does he avoid a headlong rush into mental pathology.[16]

In one sense, of course, Lewis's depiction of the barbarity of science reflects the particular concerns of postwar American literature. Like his acerbic contemporaries H. L. Mencken and Sherwood Anderson, Lewis scorned what he viewed as the domestication of industrialized society. The conventions of efficient accumulation found in popular domestic and workplace guides of the 1920s (epitomized in Stuart Chase's 1927 *The Tragedy of Waste*) portrayed any unproductive expenditure as a moral indecency.[17] But Lewis and other writers took issue with this dogmatism. As Alfred Kazin describes, these intellectuals rebelled against an America they viewed as dangerously "soft."[18] In this respect, Arrowsmith's barbarity reveals the disgust with cultured norms of civility and efficiency that Lewis shared with leading writers of the era.

The barbarity of science, however, not only points to an idealization of the "wild" generally present among 1920s intellectuals but, more significantly, to an ambivalence about emotional affect that diverges from previous instances of self-sacrifice. From the purist's devotion to the X-ray experimenter's enthusiasm, the scientist's compulsive love for science has been depicted as central to his willingness to suffer for it (or "her"). In *Arrowsmith*, by contrast, the scientist is distinguished by his lack of emotional vulnerability. Here, the essence of science is experimentation, which demands dispassionate cruelty. That perspective reveals Lewis's reliance on his collaboration with de Kruif.

As mentioned, de Kruif, who published thirteen best-selling books of his own, trained in bacteriology at Michigan and the Rockefeller Institute under the physiologist Loeb.[19] From Loeb and his colleagues, de Kruif inherited a belief in the mechanistic determination of all natural phenomena and a commitment to rigorously controlled experimentation as the appropriate means for discerning those mechanisms.[20] To acquire knowledge, he insisted, hypotheses must be tested in narrowly defined, strictly managed, reproducible experiments; performing these experiments required a certain kind of "scientific worker," one able to eradicate all emotional sensitivity. "No matter how valuable [emotions] may be in other walks of life," de Kruif asserted in *Century Magazine* shortly before he met Lewis, they are "out of place in science." The experimenter must be "judicial" and "unemotional," eliminating his feelings as he eliminates tainting external variables.[21] Factors that determine experimental results "are to be weighed and measured meticulously and coldly, without enthusiasm for one, or disdain and enmity toward another."[22] For de Kruif, scientists were "heartless persons" consumed, like all explorers, by a single incessant impulse: "to know, to discover, to find out the how of things."[23] As he wrote in 1922:

> Contrary to their own occasional pretension, and to the invariable description of them in newspapers and popular gazettes, their desire is not primarily to relieve suffering humanity. They are driven to their task by that strange instinct that gave rise to the impudent and dangerous explorations of Henry Hudson and to the unholy and heretical probings of Galileo.[24]

This insensitivity to human suffering, de Kruif insisted, distinguished scientists not just from physicians but from the masses more generally. "Humanity in general is certainly not ready for a martyrdom to the progress of knowledge," he summarized in one description of experimental science. "It is doubtful, indeed, whether the generality will ever become the prey of such a disemboweled itch for knowledge."[25]

In following de Kruif's lead, Lewis's novel disrupts certain aspects of nineteenth-century sacrificial ethics. Where previously the special sensitivity of the scientist readied him for sacrifice (unlike those allegedly insensate bodies against whom the voluntary sufferer was defined), in *Arrowsmith* it is a special insensitivity that distinguishes the "disemboweled" scientific self. Sensitivity becomes equated with civility, which is in turn feminized and racialized. The absence of emotion signals manly self-possession.

Soon after de Kruif met Lewis in 1922, the outspoken young bacteriologist quickly came to influence Lewis's conception of the new novel, acting at once as literary model, informant, and critic. Lewis affirmed his influence in numerous letters to publisher Alfred Harcourt:

> It gives me joy to inform you that De Kruif is perfection. He has not only an astonishing grasp of scientific detail; he has a philosophy behind it, and the imagination of the fiction writer. He sees, synthesizes, characters. You've sometimes said that my books are meaty; this will be much the meatiest of all—characters, places, contrasting purposes and views of life; and in all of this there's a question as to whether he won't have contributed more than I shall have.[26]

Through the figure of Gottlieb in particular, Lewis emphasizes the scientist's necessary renunciation of compassion. Gottlieb, like Loeb, has been educated in the tradition of German experimentalism and possesses "a diabolic insensibility to divine pity, to suffering humankind" (332).[27] He "would rather have people die by the right therapy than be cured by the wrong" (120). Such insensitivity, Lewis follows de Kruif in suggesting, is essential to the life sciences, which demand experimenters' cold indifference to the suffering of living subjects. As Gottlieb calmly and deftly kills a squirming guinea pig with a vial of anthrax in front of his rapt class, he lectures: "Some of you will think that it does not matter; some of you will think, like Bernard Shaw, that I am an executioner and the more monstrous because I am cool about it; and some of you will not think at all. The differences in philosophy is what makes life interesting" (36).

Nauseated by the experiment, one student lumbers home from Gottlieb's course meditating on the relation between the guinea pig's death and the "sacrifice of the martyrs" exalted in Christian theology (37). The student, who appears to lack the indifference to suffering requisite for experimental work, ultimately decides to pursue another career. Clearly, science necessitates a specific kind of emotionality, and one with its own attendant styles of embodiment.

The scientist might be "heartless" when conducting experiments, the novel notes, but he must *passionately* detest false ideas. In a speech that one Lewis scholar describes as Arrowsmith's "ritual initiation" into the world of pure science, Gottlieb informs his pupil that he must have a heart filled with hatred for

> the preachers who talk their fables . . . the anthropologists and historians who can only make guesses. . . . the ridiculous faith-healers and chiropractors . . . the doctors that want to snatch our science before it is tested and rush around hoping they heal people. . . . worse than the imbeciles who have not even heard of science, he hates pseudo-scientists, guess-scientists . . . he hates the men that are allowed in a clean kingdom like biology but know only one text-book and how to lecture to nincompoops all so popular! (267–68)[28]

Elsewhere, Gottlieb continues to advocate a hearty contempt for untruth and a staunch intolerance for those would propagate it. He encourages Arrowsmith to do "violence to all the nice correct views of science" (53) and assails him when he feels the young man has taken up "monkey-skipping and flap-doodling" in lieu of hardboiled research (303). Intellectual creativity must be pared away to make room for the destruction of false ideas: students who "wish a liddle bit to become scientists," Gottlieb tells Martin, must be seized, denounced, and taught "right away the ultimate lesson of science, which is to wait and doubt" (15).

While Gottlieb embodies the need for scientific contempt, Terry Wickett—the only other experimenter in *Arrowsmith* to escape Lewis's satirical condemnation—is equally uncivil. At their first meeting, Arrowsmith is put off by Wickett, flustered when the chemist aggressively inquires whether he intends to be "one of the polite birds that uses the Institute for social climbing" or "one of the roughnecks like me and Gottlieb" (274). Initially disgusted by Wickett's manner, Arrowsmith eventually comes to understand and mimic him, following Terry "the barbarian" to his homemade laboratory in Vermont (427). Experimental science, Lewis indicates, demands passionate, Nietzschean contempt—a hatred of all received ideas and their cowlike adherents.

This hardened, hate-filled quest for knowledge hardly invites the feminine metaphors that pervade the writings of late nineteenth-century proponents of pure science. Truth is no beguiling goddess, making revelations, as G. Stanley Hall wrote in 1894, to "those who truly love and wait upon her."[29] In *Arrowsmith*, truth is a "clean, cold, unfriendly" set of "fundamental laws" (217, 116). Lewis compares the search for these laws to the detective's snaring of criminals, the surgeon's excision of tumors, and the soldier's violent ambush of the enemy (296, 297, 301). Truth is, in short, more akin to the "terrible glory of God" than to "pleasant daily virtues" (116).[30] The reference to the "terrible glory of God" points to the tension inherent in Lewis's depiction of the scientific life. For all of

Arrowsmith's sweating, raging contempt, his stance is portrayed as essentially sublime, noble rather than brutish.

Lewis had long intended to write a book featuring a religious hero. After the commercial success of his preceding books, the biting *Babbitt* and *Main Street*, Lewis intended his next novel to quell the widespread criticism that he lacked "spiritual gifts."[31] To this end, he planned a novel about the labor movement, featuring a Christ-like leader modeled after socialist Eugene Debs.[32] As Lewis wrote to his wife Grace in 1922, "a Debs lasts; he is pure spirit; he would walk to his crucifixion with firm & quiet joy."[33] But Lewis put aside his religious novel the moment he met de Kruif. In the beleaguered bacteriologist (de Kruif had recently lost his position at the Rockefeller Institute after publishing a critique of medical greed), Lewis found a new model for his spiritual novel. As Lewis scholar James Hutchinsson notes, de Kruif was "every bit as admirable and courageous as Debs but associated with a higher social class and a more reputable and idealistic calling." Although he empathized with the blue-collar world of Debs's labor agitation, Lewis may have been squeamish about bringing his satirical style to settings with which he had no personal familiarity. In Hutchinsson's view, "Lewis obviously felt more comfortable with this class of people and more confident of being able to portray them in his fiction."[34]

While Lewis relied on de Kruif as a model for the scientific and biographical details of *Arrowsmith*'s characters, he diverged from de Kruif in emphasizing the essentially religious character of the project of science. Scientists in *Arrowsmith* are anything but the heretics described by de Kruif in 1922; for Lewis, the piety of scientific work is its foremost grace, and Max Gottlieb best represents the religious character of scientific research. Described as a "shadow" of a person, Gottlieb shuns the body to the best of his ability (13). While a professor at Winnemac, the tall, gaunt man lived in a dank, cramped cottage and rode a squeaky old bicycle to school. Although married, he wed his "thick and slow-moving and mute" wife without thought, "as he might have bought a coat or hired a housekeeper" (122, 119). Gottlieb was a man for whom "abstractions and scientific laws were more than kindly flesh" (291).[35]

Arrowsmith perceives that this neglect of or contempt for the body is the sign of Gottlieb's greatness, his "priestly" asceticism (38). He condemns his "idiot" classmates for failing to recognize "what a man like Max Gottlieb means" (31). "You think Gottlieb isn't religious," young Martin argues with a classmate. "Why, his just being in a lab is a prayer" (31). In a way, the classmates' failure to appreciate Gottlieb's manhood is understandable since that manhood stems, paradoxically, from his diminution of the body. When Arrowsmith first sees Gottlieb, the professor appears as

> a tall figure, ascetic, self-contained, apart. His swart cheeks were gaunt, his nose high-bridged and thin. He did not hurry, like the belated homebodies.

He was unconscious of the world. He looked at Martin and through him; he moved away, muttering to himself, his shoulders stooped, his long hands clasped behind him. He was lost in the shadows, himself a shadow.

He had worn the threadbare top-coat of a poor professor, yet Martin remembered him as wrapped in a black velvet cape with a silver star arrogant on his breast (13).

Gottlieb's power, in other words, arises from his renunciation of the world of flesh. Martin's wife Leora confirms this power. Referring to Gottlieb as "the greatest man" she'd ever seen, she informs her husband that the emaciated scientist is "the first man I ever laid eyes on that I'd leave you for, if he wanted me'" (118). Manliness increases with every demonstration of bodily denial.

This ascetic practice is no self-effacing piety, however. As the barbarity of Arrowsmith's actions make clear, transcendence demands a vigorous, violent scorn for civilization and its trappings. From the "sacrifice" of Gottlieb's guinea pig to the aggressive tones of Martin's "prayer of the scientist" (269), the religious ascetic and the "contemptuous acolyte of science" are said to share a hardened quest for a higher truth. Terry "the barbarian" exemplifies this hardness. Wickett's jibes, as Arrowsmith realizes, "made up the haircloth robe wherewith he defended a devotion to such holy work as no cowled monk ever knew" (410).

The climactic episode of the novel, when Arrowsmith and his entourage travel to the plague-stricken island of St. Hubert to test his phage, condenses and reveals the tension between a barbaric commitment to truth and a civilized humanitarianism.[36] It also demonstrates how racialized histories of voluntarism allow the denial of life-saving medication to the island's inhabitants to be portrayed as noble: unlike the colonial subjects of St. Hubert, the fully self-possessed visitor might (and must) willfully surrender his heart along with his body. Before he departs for the island, Gottlieb tells his protégé that only refusing the human temptation toward pity can preserve his status as a scientist. Martin must resist the cry of his "own good kind heart," lest his humanity "spoil" the experiment (338). Humanitarianism is not in itself a contemptible value, Gottlieb concedes; but above all, foremost and first, there "must be knowledge," acquired only through a cold willingness to allow some human beings to go untreated as an experimental control (338). Rather than refuting his instinct for pity entirely, Gottlieb suggests, Arrowsmith must allow his pity for imagined successive generations to douse his eagerness to treat the sick and suffering of today. "Let nothing," Gottlieb repeats firmly, "neither beautiful pity nor fear of your own death, keep you from making this plague experiment complete" (339).

Yet once he witnesses the tormented deaths of so many of the islanders, Martin wavers in his desire for experimental completeness. In his moment of weakness, "he regained the picture of Gottlieb's sunken, demanding eyes; and he swore that he would not yield to a compassion which in the end would make all compassion futile" (359). The question for Arrowsmith, as for the "volun-taries" of Emerson's poem, was, Which suffering was worthwhile and on what basis? With what conviction might the reason for the exchange be assured? As Arrowsmith prepared to justify his plan to the island's board of governance, he summoned the image of his mentor for sustenance. "Beside him stood Max Gottlieb, and in Gottlieb's power he reverently sought to explain that mankind has ever given up eventual greatness because some crisis, some war or election or loyalty to a Messiah which at the moment seemed weighty, has choked the patient search for truth" (361).

Despite the calming assurances of the biblical Gottlieb, Arrowsmith's plan to leave some of the colonial subjects untreated triggers an enraged protest from the island's governors. The emotive civility of colonialism is set in opposi-tion to the barbarity of Arrowsmith's refusal to administer the treatment uni-versally. Colonial administrator Sir Robert Fairlamb bellows at Arrowsmith's proposal:

> "Young man, if I were commanding a division at the front, with a dud show, an awful show, going on, and a War Office clerk asked me to risk the whole thing to try out some precious little invention of his own, can you imagine what I'd answer? There isn't much I can do now—these doc-tor Johnnies have taken everything out of my hands—but as far as possi-ble I shall certainly prevent you Yankee vivisectionists from coming in and using us as a lot of . . . sanguinary corpses." (360)

Apologizing to his wife, "Lady Fairlamb," for his use of strong language and maintaining the cordial "Sir" at the end of his address, Fairlamb epitomizes the mannered civility that Martin—emboldened by the image of Gottlieb—rejects. Arrowsmith invokes the safety of his racial privilege to boast of his resistance to the people who are trying to compel him to administer the phage: "Nothing can make me do it, not if they tried to lynch me" (375). When Leora dies on the is-land, however, he falls apart. Soon he is injecting everyone, taking "a bitter sat-isfaction in throwing away all his significance, in helping to wreck his own purposes" (376). Destroying the opposition to practical utility that gave his work meaning, Arrowsmith feels that he is a "traitor to Gottlieb and all that Gottlieb represented." The more the people on the island "shouted his glory, the more he thought about what unknown, tight-minded scientists in distant laboratories would say of a man who had his chance and cast it away. The more they called him the giver of life, the more he felt himself disgraced" (381).

What the panoptic gaze of these unknown scientists suggests, Lewis im-plies, is an affinity between the concerns of experimental scientists and cowled monks: a shared devotion to purposeful impracticality, to the ceaseless quest for an ever-receding truth. Like polar explorers dying to reach an obscure mathematical point or radiologists scorching their fingers in an effort to discover the source of the X-ray "burns," Lewis's scientist lunges into his search for "the underneath principle" that he well knows he can never obtain (53).[37] Although ultimately futile, this incessant search for truth is not unpleasant: like Walter J. Dodd or G. Stanley Hall, Arrowsmith is never so happy as when facing insurmountable new "mountain-passes of work" (301). He "beautifully and excitedly fail[s]" in his experiments (247). He *must* fail, in the end, since his goal—his ennobling, purifying goal—is to accede to a fugitive, inexhaustible truth. (Lewis toyed with titling the book "The Merry Death.")[38] Reveling in the ceaselessness of his quest, Arrowsmith contentedly studies all night to prepare for the "work whose end is satisfying because there is never an end" (287). As a grown man in Vermont, Arrowsmith looks forward to the failure of his investigations of quinine derivatives; as a boy, he "blissfully" looked for, but never managed to find, "the Why that made everything so" (54).

From one perspective, Lewis's emphasis on the eternal failure of Arrowsmith's investigations seems perfectly in keeping with the imperatives of a growing consumer society. Science, like beauty, might be seen as the kind of luxury, of conspicuous consumption, that ultimately affirms the norm of measured, temperate productivity. One member of the American Association for the Advancement of Science affirmed this view in 1920, suggesting that scientific research, like "artistic creation" or "philosophic speculation," is a "reservation" for the mind: "great world parks to which man must resort to escape from the deadening, overspecializing routine of his habits, mores and occupations and enjoy veritable creative holidays of the spirit."[39] Here, science's uselessness offers a respite that ultimately allows the machinery of civilization, including "man" himself, to spin on smoothly.

Lewis, however, offers a more profound critique of the objectives of productivity, allowing Arrowsmith to triumph as a heroic figure only after disavowing consummation and fulfillment altogether. Emphasizing this point, Lewis matches the idealized futility of Arrowsmith's scientific research to the unproductivity of his marriages. Arrowsmith defies reproductive norms by abandoning his only child, casually throwing off paternity in favor of an adult relationship with Terry Wickett. But he also subverts the norms of monogamous, reproductive heterosexuality in his relationships with women. He terminates his youthful romance with the white-limbed, elegantly dressed Madeline Fox for the "vulgar, unreticent" Leora Tozer (56). While he longs physically "for the girl Leora," Arrowsmith entrusts his emotional life to "another, sexless Leora," one who "with her boyish nod or an occasional word . . . encouraged him to confidence" (70–71). In "her indolence, her indifference to decoration and good fame," Leora "was neither woman nor wife but only her own self" (397).

Like Leora, here described as alternately boyish and sexless, the two other women in Arrowsmith's life also appear attractively defeminized. Orchid Picker-baugh, the nineteen-year-old daughter of Martin's boss in Nautilus, Iowa, was never so appealing to Martin as when playing in the snow, dressed in the "roughness of tweeds, eyes fearless, cheeks brilliant as she brushed the coating of wet snow from them, flying legs of a slim boy, shoulders adorable in their pre-tense of sturdy boyishness" (212). Similarly, Joyce Lanyon, the woman who be-comes Arrowsmith's second wife, strikes him as attractive at the same moment she appears to be his physical double: "she was his twin; she was his self en-chanted" (368). Drawn by androgynous or boyish female figures, Martin quietly secedes from the norms of reproductive heterosexuality, affixing his affections to Wickett and Gottlieb (33, 314) and his lust to an infinitely receding truth.

Exemplifying the rejection of softness prevalent among literary intellectu-als of the 1920s, Lewis presents the feminine as the antithesis of science throughout the novel. Experimentation, the emblematic activity of the brutal assault on truth, only acquires feminine attributes when ill-conceived or poorly executed. A professor at Arrowsmith's medical school, for instance, conducts this sort of effeminate research: "When Robertshaw chirped about fussy little experiments, standard experiments, maiden-aunt experiments, Martin was restless" (22). In one of the rare passages where science itself is feminized, the speaker is no researcher at all but an eminent pathologist who abandoned the "tough" work of experimentation for the decorative work of institutional ad-ministration. The metaphorically emasculated researcher, Dr. A. DeWitt Tubbs, bubbles to Martin, "'I'm afraid I'm not the moral man that I pose as being in public! Here I am married to executive procedure, and still I hanker for my first love, Milady Science!'" This confession of scientific desire strikes Martin as disingenuous, however, since he notices that Tubbs's laboratory bench shows no sign of recent use (274). The feminization of science in Tubbs's statement only acts to accentuate his detumescence as a researcher.

The only explicitly feminized object in the McGurk Institute's laboratory space, an enormously expensive and largely ornamental centrifuge nicknamed "Gladys," happens also to be the only notably dysfunctional object in the novel. Gladys fails to separate materials and frequently flings the contents of test tubes out into the laboratory. Uncontained and uncontainable, the leaky centrifuge is finally dismissed from the McGurk for "her sluttish ways," and Martin and Terry get "bountifully drunk" to celebrate "her" departure (418). The sale of the centrifuge purportedly purifies the institute (315). But like the psychoanalytical repressed, Gladys returns. Joyce, Arrowsmith's second wife, acquires the expen-sive apparatus when constructing a laboratory over their garage for Martin's use. Her proud acquisition of the discarded machine confirms Arrowsmith's perception of women's scientific ineptitude. The feminine—whether evident in the cloying affection of Joyce Lanyon or the mechanical incompetence of Gladys—poses a continual obstacle to Arrowsmith's science.

Femininity is also tied to domesticity, which threatens to tame barbaric, unproductive science. Arrowsmith's laboratory technician, while competent, cannot become a real scientist since he favors "six hours' daily sleep and sometimes seeing his wife and children" (324). Invariably, the temptations pulling the scientific hero away from his barbarous pursuit of science are cloaked in similar domestic imagery. Representatives of the home appear throughout the narrative to try to convince Arrowsmith of the value of utility and productivity. Dean T.J.H. Silva, for instance, scolds Arrowsmith for having an irrational infatuation with knowledge for its own sake. Echoing an earlier article by de Kruif, "Dad" Silva tells Martin:

> "It's all very fine, this business of pure research: seeking the truth, unhampered by commercialism or fame-chasing. Getting to the bottom. Ignoring consequences and practical uses. But do you realize if you carry that idea far enough, a man could justify himself for doing nothing but count the cobblestones on Warehouse Avenue—yes, and justify himself for torturing people just to see how they screamed." (117)[40]

The father reprimands his wayward son for scattering intellectual seed. Work must be useful, he insists; energies must be spent wisely and fruitfully. Despite Dad's best efforts, Arrowsmith maintains his sense of "higher" value. When the dean's efforts to lure Arrowsmith toward useful research are thwarted by Martin's stubborn attachment to research, "Dad" Silva enlists the aid of that other key representative of the home: Mrs. Arrowsmith. Insisting that Martin should be "passionate on behalf of mankind," he admonishes Leora, " 'You must keep him at it, my dear, and not let the world lose the benefit of his passion' " (118). At the other end of the spectrum, Gottlieb, exemplar of authentic science, repeatedly criticizes marriage in front of Arrowsmith, sneering at "dese merry vedding or jail bells" (105).[41] Arrowsmith himself waffles between the world of sensible domesticity and unreasonable scientific extravagance. After marrying for the first time, for example, he vows to take up useful, lucrative habits. "He had Leora now, forever. For her, he must be sensible. He would return to work, and be Practical. Gottlieb's ideals of science? Laboratories? Research? Rot!" (103). When Arrowsmith at last renounces Joyce and commits himself to pursuing science in the wilderness, only "maverick and undomestic researchers" are welcome (427).

Despite the novel's apparent antithesis between unscientific, feminine domesticity and scientific, masculine barbarity, it is worth pointing out that women characters themselves do little to disable scientific research. While Gottlieb curses the prison of marriage, for instance, it is his "mute" wife who first enables his presence in the lab and his daughter Miriam who later supports his science emotionally, financially, and practically. Leora goes still further, resisting efforts to turn her wild husband into a polite and respectable citizen and steering Arrowsmith toward the hard, cold life of research as he reluctantly backs away. "You belong in a laboratory," she insists when he considers "settling down" to a useful occupation: "finding out things, not advertising them. . . . Are you going on for the rest of your life, stumbling into respectability and having to be dug out again? Will you never learn you're a barbarian?" (210) While the novel's male characters contrast femininity and authentic research, the actual women in the story in fact protect and promote their research. The salient struggle for self-definition, then, lies within the powerful men at the center of the story—in their own persistent efforts to mold themselves to the demands of science.

By feminizing the productivity of civilization and masculinizing an unproductive barbarism, Lewis recoups Arrowsmith's blissful pursuit of an unattainable goal as testimony to his manliness, his scientificity, and his moral virtue. Far from perverse, his rejection of production and reproduction appear eminently laudable, the sign of individual perseverance and integrity. What de Kruif failed to realize when he first read the *Arrowsmith* manuscript is the degree to which the rejection of compensation—the *loss* at the heart of this exchange with science—produces Arrowsmith's heroic status. While Lewis's biographer discards the novel's conclusion as "a little fantastic," Arrowsmith's happy expectation of failure actually represents his ascendance as a self-sacrificing scientist. What Lewis so skillfully reveals is a key tension in standards of sacrifice: Gottlieb is at once emaciated and arrogant; Wickett is at once barbaric and eminently admirable; Arrowsmith's life is at once squalid and transcendent. While the real-life scientists we have seen in preceding chapters struggle to come to terms with the contradictions of sacrifice, Lewis's fictional scientist inhabits them triumphantly, lolling happily on a placid lake.

Epilogue

The Ends of Sacrifice

By maintaining self-sacrifice as heroic yet inverting the assumption of privileged civility that conditioned the rise of voluntary suffering, Lewis's *Arrowsmith* displays the fruition of late nineteenth-century sacrificial ideals. By the 1920s, the figure of the voluntarily suffering scientist was the subject of parodic exaggeration and even more overt critique. Popular articles with titles such as "Scientists at Play" and "The Fun of Being a Scientist" suggest that preoccupation with self-sacrifice had become as likely to be a source of amusement or embarrassment as of reverence and respect.[1] A survey of popular periodical literature through 1932 reveals a spike in articles invoking sacrifice or self-sacrifice in the 1910s and then a steep decline after 1923. Thus, even as Lewis's acclaimed novel drew a wide readership, explicit references to self-sacrifice were on the wane.

The reasons for this movement were surely overdetermined. It may be that the fresh trauma of world war dimmed the nostalgic glow that had suffused the coupling of progress and suffering since the 1870s. Perhaps the increasingly widespread use of anesthetics amplified mechanistic understandings of pain, thereby challenging the moral significance of suffering. Perhaps continued activism against human and animal experimentation troubled enthusiastic tributes to the role of injury and death in science. Perhaps the carnival atmosphere surrounding the 1925 Scopes "monkey" trial hushed the use of inflated religious metaphor in scientific practice.[2]

Or, perhaps more accurately, we might say that the waning rhetoric of self-sacrifice in the mid-1920s indicates that the ethic had been thoroughly absorbed into everyday life. References to the "martyrs of science" continued to crop up from time to time throughout the twentieth century. Each research-related death of a prominent scientist—such as Rockefeller Institute researcher Hideyo Noguchi (who succumbed to yellow fever in 1928) or Marie Curie (whose death from leukemia in 1934 was widely attributed to her work with radium)—

generated a fresh opportunity to reinvent the tradition of voluntary suffering.[3] But the declining persistence of explicit references to self-sacrifice may suggest assimilation rather than extinction of these values. Even as explicit religious and moral language fell out of fashion, understandings of the scientist as a willing sufferer, and of science as a field that demands unusual amounts of pain, continued to hold power.

As we have seen, this power was predicated on a paradoxical, exclusionary "self of no self." Cook's and Peary's frozen noses and aching feet, for example, testified to a form of self-possession not allowed to Eskimo and Negro assistants. X-ray experimenters Kassabian, Glidden, and Leonard were hailed as willing martyrs to science, while laborers and animals subjected to equivalent amounts of X radiation were not. To consecrate oneself to truth in nineteenth-century America required a certain kind of socially constituted self. Without this willful self, one could hardly take up the peculiar privilege of voluntary suffering.

That self-sacrifice for science reflected and reproduced larger social divisions is just one lesson to be learned by connecting histories of proprietal selfhood to those of scientific practice. The more subtle and interesting observation is that even privileged subjects described themselves as bound to the demands of science. For these scientists, as for others formed by the legal and conceptual traditions of possessive individualism, freedom implied the liberty to determine one's own subjection. Among individuals conceived as proprietors, action was necessarily governed by judgments of comparative value. To be a reasonable agent in the modern market system of post-Emancipation America—indeed, to be *human*—is to act as a self-interested possessor, to calculate one's investments.[4]

These calculations, we have also seen, generate ongoing uncertainty. The issue at stake for the scientists discussed in this book was not merely the discrepant value of the various goods surrendered or obtained but whether sacrifice was properly considered an act of exchange at all. For some, self-sacrifice for science fit handily into a larger picture of compensatory exchange, a vision in which (as Emerson argued) "nothing is given, all things are sold." X-ray experimenter Emil Grubbé exemplified the reciprocal vision of self-sacrifice when he declared that each secret obtained in "the vineyards of science" exacted a corresponding personal cost. The more valuable the advancement, he proposed, the higher the price paid by the individual scientist. Echoing the original social contract imagined by liberal philosophers, this understanding of sacrificial science presumed that the investigator pursued ends of obvious equivalence to his relinquishment. Indeed, the perceived value of the gain *must* be roughly equivalent to individual expenditure if the reasonable man were to partake in it. As Georg Simmel, one of the era's most important theorists of sacrifice, argued in 1900, unless a balance appeared between the value of the good lost and of the good received, "no agent would consummate the exchange."[5] "No one in his right mind," he repeated in 1907, "would forego value without receiving for it at least an equal value."[6] Free agents could hardly do otherwise than to judiciously gauge their self-interest without calling into question the very meanings of freedom and agency.

Yet for others, such deliberate, self-interested reciprocity opposed the elevating purpose of voluntary suffering. To seek knowledge of obvious equivalence to suffering would invalidate the "sacrificial" aspect of the activity. The subjects of this book often aligned themselves with this more asymmetrical view of the relation between scientist and science. For them, calculations of personal compensation contradicted the unbridled, excessive devotion that distinguished the true scientist. Rowland, Remsen, and Hall insisted that purity occasioned an inexhaustible longing for a perpetually receding truth. MacMillan proposed that explorers surrendered their noses and toes for a spot of ice "barren and desolate beyond all imagination." Dodd and the other "martyrs" persevered enthusiastically in their lethal experiments, even while acknowledging the ray's destructiveness. Arrowsmith renounced family, wealth, and social status for a life of inquiry in the Vermont woods, triumphantly predicting that his efforts would fail. In these varied examples, an expressed disdain for calculations of utility was not a secondary or epiphenomenal feature of sacrifice for science: it was its defining attribute.

This is not to say that self-sacrifice implied simple waste for any of these real or fictitious scientists. Sacrifice entailed a rejection of practical utility but was not to be confused with hapless, degenerate squander (although Martin Arrowsmith veered dangerously close to this line). "Merely to give up a good," asserted one university professor, "is to waste goodness, not to make a sacrifice."[7] Suffering pursued for its own pleasure, as in cases of masochism, signaled only perversion; suffering endured for lack of an alternative, as in conditions of slavery, signaled only the debasements of bondage. Nevertheless, even the most modest sacrifice had to exceed the measured circularity of the consensual exchange. To be viewed as a true gift, the offering must be at once purposeful (a gift extended for an end that "lies beyond the agent making the sacrifice") and inescapable, resting somewhere in the realm of unreasonable, impassioned compulsion.[8]

If valorization of the voluntarily suffering self has yet to end, it is in part because the uncertainties underlying this self—uncertainties inherent to the modern liberal subject—persist unabated. Given the constitutive ambiguities of proprietal selfhood, it is not surprising to find the language of compulsion and captivity used in contemporary descriptions of the relation between scientist and science. When, in 2003, Alan Lightman told readers of the *New York Times* that "the real reason a scientist does science" is because "the scientist *must*," he echoed themes of willing bondage dating back the original theorists of social contract. When we read that "love" for investigating the unknown is at once "a gift filled with beauty and not given to everyone" and a "burden because the call is unrelenting and can drown out the rest of life," it is clear that the ambiguities and exclusions of the proprietal self are with us still.[9]

Indeed, as the imperatives of transnational capitalism summon this property in ever more intricate ways (rendering alienable not only our capacities for love and labor but now also genes and cell lines), we might predict a resurgence of longing for willing sacrifice—for impassioned subjection to "knowledge for its own sake." Given yearnings for some nonalienated, if not sacred, sense of self, for more profound experiences of "belonging," is it any wonder that the nation's preeminent newspaper waxed nostalgic about the "good old days, when science students suffered"?[10] As we continue to look to science for emancipation and to suffering for elevation, it may prove helpful to recall the suffering for science chronicled here, remembering the relational and contingent identity of the sacrificial self. Revisiting past definitions of reasonable suffering leads us to recognize not only how we, too, are still pious, but also how we, too, indulge the peculiar pleasures of pain.

ACKNOWLEDGMENTS

The congruities of researching and writing about people who suffer for knowledge did not escape my attention. Many people helped me through the long and sometimes painful process of completing this book. My debts to them are innumerable.

I am grateful to the five different academic communities that wagered financial resources on this project: the Program in Science, Technology, and Society at the Massachusetts Institute of Technology (MIT); the Department of History of Science and Technology at Kungl Tekniska Högskolan (KTH) in Stockholm; the Dibner Institute for the History of Science and Technology in Cambridge, Massachusetts; the Centre for the History of Science, Technology, and Medicine at the University of Manchester; and the Program in Women and Gender Studies at Bates College. I also thank colleagues at Colby College, KTH, Leeds University, Manchester University, MIT, Rensselaer Polytechnic Institute, the University of California at San Diego, and the University of New Hampshire for opportunities to present previous versions of this material. Had I the capacity to incorporate more fully those audiences' many thoughtful comments, this surely would be a better book. My gratitude goes as well to the many librarians and archivists who aided this project over the years, particularly Heidi Herr at Johns Hopkins University; Helen Samuels at MIT; John Waggener at the University of Wyoming; and Tom Hayward, Perrin Lumbert, and Chris Schiff at Bates. For permission to reproduce unpublished or previously published material, I thank the American Heritage Center at the University of Wyoming, the Archives and Special Collections Department of Clark University, the Boston Medical Library in the Francis A. Countway Library of Medicine, the George J. Mitchell Department of Special Collections and Archives at Bowdoin College, the Johns Hopkins University Press, and the Ferdinand Hamburger Archives at Johns Hopkins University.

My thinking about science, American history, and words and the worlds to which they refer owes more to Leo Marx than to anyone else. Among the countless ways in which Leo contributed to the making of this book, perhaps most crucial was his insistence that I write, as he put it, from my "own deepest convictions." Leo's serious, patient commitment to bringing those convictions to light not only sustained my completion of this study but also emboldened the rest of my endeavors. I extend to him my most heartfelt appreciation. Deborah Fitzgerald and Evelynn Hammonds joined Leo in supervising the 1998 dissertation that grew into this book; I thank them for their incisive suggestions on that version of the project and for their continued guidance.

Other teachers also inspired this effort. I am especially indebted to those responsible for the remarkable education I received in the California public school system. In Lakeside, those people include Maureen Begley, Greg Carlson, Jan Coulsby, and Jerome Lipetzky; and in Santa Cruz, they include Jennifer Anderson, Chris Connery, Donna Haraway, Paul Niebanck, and Dan Scripture. I am daily grateful for all they have made possible for me.

At the other end of the country, members of two different writing groups helped me to gestate this book over the years: in Massachusetts, Vitka Eisen, Paula Hooper, Angela Radan, and Barbara Schulman; and in Maine, Lisa Botshon, Monica Chiu, Robin Hackett, Melinda Plastas, Eve Raimon, and Siobhan Senier. Audra Wolfe, my editor at Rutgers, helped move the manuscript from its pupal form with her enthusiastic response to an early essay in *American Quarterly*; Dawn Potter's superb copyediting rescued the book from more numerous infelicities. I owe a great deal as well to two anonymous reviewers for their astute critiques of early material submitted to the Press. Laura Briggs reviewed the completed manuscript, and her perceptive and heartening comments greatly assisted my final revisions. I also thank the many colleagues and students at Bates whose kind inquiries helped nudge this book out the door. I am especially grateful to Rachel Austin and Erica Rand, whose attentive readings of the entire manuscript improved it immeasurably, and to Lorelei Purrington and Regan Richards, whose expert assistance and good cheer were vital as deadlines approached.

The influence of Jill Hopkins Herzig, my mother, hiking partner, and lifelong intellectual companion, appears everywhere in these pages. Her way of being in the world continues to establish for me the standard by which to gauge "how one should live." Joel, Micah, and Rachel Herzig brought mirth and camaraderie to this effort as to all others. Jenny Reardon, my interlocutor on all subjects, has consistently sought to understand the connections between this study and the rest of my life, and for that loving effort in particular I am thankful.

For generative conversations, careful readings, and general encouragement, I also thank Jon Adler, Jon Agar, Kiran Asher, Jim Austin, Christine Baldwin, Bennett Battaile, Connie Battaile, Gordon Battaile, Jenny Beckman, Danielle Benedict, Roberta Bivins, Curtis Bohlen, Geoffrey Cantor, Heidi Chirayath, Greg Clancey, Stephen Collier, Bill Corlett, Jim Daly, Lorraine Daston, Elizabeth Eames, Michael Fischer, Dan Freeman, Mats Fridlund, Jan Golinski, Penelope Gouk, Leslie Hill, Jeff Hughes, Margaret Imber, Arne Kaijser, Thomas Kaiserfeld, Evelyn Fox Keller, Sasha Keller, Chris Kelty, Sharon Kinsman, Andy Lakoff, Hannah Landecker, Svante Lindqvist, Tina Malcolmson, Lisa Maurizio, Andy Martin,

Jane Marx, Don McCarthy, Ted Metcalfe, David Mindell, Marvin Mindell, Phyllis Mindell, Jennifer Mnookin, Silver Moore-Leamon, Naomi Oreskes, John Pickstone, Buck Pelkey, Jill Reich, Harriet Ritvo, Michael Robinson, Michael Sargent, Lena Schnell, Heinrich Schwarz, Steven Shapin, Bonnie Shulman, Katie Shultz, Merritt Roe Smith, John Staudenmaier, John Tresch, Alison Vander Zanden, Debbie Weinstein, Anita White, and Nina Wormbs.

Since this book is, in many ways, a meditation on loss, it is fitting to dedicate it to Sidney J. Herzig and Norman O. Brown, whose deaths (one sudden and shocking, one long anticipated and discussed) merely served to accentuate their pervasive, abiding influence on my life and thought. To the extent that it is mine to offer, this book is for them.

NOTES

PREFACE

1. See, for example, Scott LaFee, "Perish the Thought: A Life in Science Some-times Becomes a Death, Too," *San Diego Union-Tribune,* 5 December 2003, pp. F1, F4; J. M. Hirsch, "Scientists Stunned by Fatal Mercury Poisoning of Colleague," *San Diego Union-Tribune,* 11 June 1997, p. A11; Carey Goldberg, "Colleagues Vow to Learn From Chemist's Death," *New York Times,* 3 October 1997, p. A7; Rick Bragg, "A Drop of Monkey Virus Kills a Researcher in 6 Weeks," *New York Times,* 14 December 1997, p. 29.
2. David Hochman, "Buried Alive and Loving It," *National Geographic Adventure* (December 2002–January 2003): 51.
3. "Group's Members Say They'll Risk Lives in AIDS Vaccine Tests," *Los Angeles Times,* 22 September 1997, p. A11.
4. Belinda J. Wagner, "A Scientist in Search of Balance," *Sojourner* 22 (January 1997): 17.
5. Nick Arnold, *Suffering Scientists* (London: Scholastic, 2000), 6, 224.
6. Alan Lightman, "Spellbound by the Eternal Riddle, Scientists Revel in Their Captivity," *New York Times,* 11 November 2003, p. D4.

INTRODUCTION   TRUTH AT ANY PRICE

1. Stubbins Ffirth, *A Treatise on Malignant Fever; With an Attempt to Prove Its Non-Contagious Nature* (Philadelphia: B. Graves, 1804), 54.
2. Ibid., 56.
3. George M. Sternberg, "Transmission of Yellow Fever by the Mosquito," *Journal of Social Science* 39 (1 November 1901): 84–99. For a list of experimental subjects, see Aristides Agramonte, "Report of Bacteriological Investigations upon Yellow Fever," *Medical News* 76 (10 February 1900): 206–7; and his "The Inside History of a Great Medical Discovery," *Scientific Monthly* 1 (1915): 209–37. The Reed experiments are now among the most famous trials in the history of U.S. (and Cuban) science. I here rely on the narratives found in U.S. accounts not in order to reproduce them uncritically but to highlight some of their rhetorical conventions. François Delaporte expertly discusses the divergence of American and Cuban historians' accounts of the Reed experiments in *The History of Yellow Fever: An Essay on the Birth of Tropical Medicine,* trans. Arthur Goldhammer (Cambridge, Mass.: MIT Press, 1991). On changes in representations of the Reed experiments over time, see Susan E. Lederer, *Subjected to Science: Human Experimentation in America before the Second World War* (Baltimore: Johns Hopkins University Press, 1995), 19–23, 131–36. Of the voluminous secondary literature on yellow fever, see Margaret Warner, "Hunting the Yellow Fever Germ: The Principle and Practice of Etiological Proof in Late Nineteenth-Century America," *Bulletin of the History of Medicine*

59 (1985): 361–82; Nancy Stepan, "The Interplay between Socio-Economic Factors and Medical Science: Yellow Fever Research, Cuba and the United States," *Social Studies of Science* 8 (November 1978): 397–423; and William B. Bean, "Walter Reed and the Ordeal of Human Experiments," *Bulletin of the History of Medicine* 51 (1977): 75–92.

4. On Reed's role in the experiments, see Lawrence K. Altman, "The Myth of Walter Reed," in *Who Goes First? The Story of Self-Experimentation in Medicine* (New York: Random House, 1987), chap. 6.

5. Sidney Howard, *Yellow Jack* (New York: Harcourt Brace, 1934); Howard's screenplay was the basis for the 1938 film *Yellow Jack*, directed by George B. Seitz.

6. Paul de Kruif, *Microbe Hunters* (New York: Pocket Books, 1926), 319.

7. Sternberg, "Transmission of Yellow Fever," 91.

8. Walter Reed, "The Propagation of Yellow Fever: Observations Based on Recent Researches," *Medical Record* 60 (10 August 1901): 201. As Altman reports, the description of Lazear's self-experiments was altered in some reports so that his widow would not have to forfeit her life insurance benefits; see *Who Goes First?*, 150.

9. de Kruif, *Microbe Hunters*, 311.

10. Altman asserts that Ffirth "left no clues" as to why he chose himself to be the subject of the experiments: "'It is unnecessary to trouble the public with my motives,' he wrote, 'and as my reasons would not affect the usefulness or duration of this work, I shall decline giving any'" (Ffirth, quoted by Altman, *Who Goes First?*, 132). In this passage from his dissertation, Ffirth is in fact discussing his motives for publishing (not for experimenting), yet the essence of Altman's point remains: Ffirth was remarkably silent on the subject of self-experimentation.

11. Ffirth, *On Malignant Fever*, 46–47.

12. Ibid., 52.

13. "Self-Sacrifice for the Sake of Science," *Current Literature* 31 (October 1901): 385.

14. "The Future of American Science," *Science* 1 (1883): 3.

15. George Bruce Halsted, "The Culture Given by Science," *Science* 4 (3 July 1896): 12.

16. Michael Pupin, *From Immigrant to Inventor* (New York: Scribner, 1923), 378.

17. Edwin Emory Slosson, "The Relative Value of Life and Learning," *Independent*, 12 December 1895, p. 7.

18. P.G.W. Glare, ed., *Oxford Latin Dictionary* (New York: Oxford University Press, 1997), 1674; Norman O. Brown, *Apocalypse and/or Metamorphosis* (Berkeley: University of California Press, 1991), 197.

19. Lederer, *Subjected to Science*, 20–24.

20. "Society Proceedings: Third Pan-American Medical Congress Held at Havana, Cuba, February 4–7, 1901," *Medical News* 78 (16 February 1901): 283.

21. Lederer, *Subjected to Science*, 21, 132–35. For discussion of Lazear's role as subordinate to Reed, see Delaporte, *History of Yellow Fever*, 98.

22. Amy Dru Stanley, *From Bondage to Contract: Wage Labor, Marriage, and the Market in the Age of Slave Emancipation* (New York: Cambridge University Press, 1998), x. Of the massive literature on these post-Reconstruction changes, of greatest influence to this study are Eric Foner, "The Meaning of Freedom in the Age of Emancipation," *Journal of American History* 81 (September 1994): 435–60; Foner, *The Story of American Freedom* (New York: Norton, 1998); Saidiya V. Hartmann, *Scenes of Subjection: Terror, Slavery, and Self-Making in Nineteenth-Century America* (New York: Oxford University Press, 1997); and Stanley, *From Bondage to Contract*.

23. J. B. Baillie, "Self-Sacrifice," *Hibbert Journal* 12 (January 1914): 260.

24. This periodization is based on close review of collections of the *American Journal of Sci-*

ence and Arts, Science, Proceedings of the American Philosophical Society, Atlantic Monthly, Harper's, Popular Science Monthly, and *Putnam's Monthly,* as well as an electronic search of the American Periodicals Series Online, which surveys 1,100 periodicals published between 1740 and 1900.

25. Within the academic field of American studies, it has become somewhat unfashionable to speak of "America" as though it represented a coherent and distinctive object of study (see Alan Wolfe, "Anti-American Studies," *New Republic,* 10 February 2003, pp. 25–32). While I am sympathetic to many of these critiques, I continue to employ the term here because period commentators used it in consequential ways. One task of the book is to convey how the post–Civil War consolidation of America influenced the imagination of science as transcendent and unified.

26. See Robert V. Bruce, *The Launching of Modern American Science, 1846–1876* (Ithaca, N.Y.: Cornell University Press, 1987), 32–35; Nina Baym, *American Women of Letters and the Nineteenth-Century Sciences: Styles of Affiliation* (New Brunswick, N.J.: Rutgers University Press, 2002), 17; and Susantha Goonatilake, "Modern Science and the Periphery: The Characteristics of Dependent Knowledge," in *The "Racial" Economy of Science: Toward a Democratic Future,* ed. Sandra Harding (Bloomington: Indiana University Press, 1993), 259–267.

27. On Culin, see Woody Register, "Everyday Peter Pans: Work, Manhood, and Consumption in Urban American, 1900–1930," in *Boys and Their Toys? Masculinity, Class, and Technology in America,* ed. Roger Horowitz (New York: Routledge, 2001), 211ff. On the contrast between nineteenth-century themes of suffering and eighteenth-century themes of pleasure, see Lorraine Daston, "Before Vocation: Science As Work," paper delivered at the University of California, Los Angeles, 17 May 2003.

28. Luce Irigaray, *Speculum of the Other Woman,* trans. Gillian C. Gill (Ithaca, N.Y.: Cornell University Press, 1985), 278ff; Page duBois, *Torture and Truth* (New York: Routledge, 1991). For divergent readings of Aeschylus's famous phrase, see *Oresteia: Agamemnon, The Libation Bearers, The Eumenides,* trans. Richmond Lattimore (Chicago: University of Chicago Press, 1953), line 178; and *The Oresteia,* trans. Hugh Lloyd-Jones (Berkeley: University of California Press, 1993), 38, note 176. At stake in this divergence is whether acsesis implied suffering for the Greeks or whether Christianity introduced pain to received practices of self-formation.

29. Leora Batnitzky, "On the Suffering of God's Chosen: Christian Views in Jewish Terms," in *Christianity in Jewish Terms,* ed. Tikva Frymer-Kensky et al. (Boulder, Colo.: Westview, 2000), 203–20; David Kraemer, *Responses to Suffering in Classical Rabbinic Literature* (New York: Oxford University Press, 1995).

30. Judith Perkins, *The Suffering Self: Pain and Narrative Representation in the Early Christian Era* (London: Routledge, 1995); Mitchell B. Merback, *The Thief, the Cross, and the Wheel: Pain and the Spectacle of Punishment in Medieval and Renaissance Europe* (London: Reaktion, 1999); Susan Sontag, "The Artist As Exemplary Sufferer," in *Against Interpretation and Other Essays* (New York: Farrar, Straus, and Giroux, 1966); Daniel Boyarin, *Dying for God: Martyrdom and the Making of Christianity and Judaism* (Stanford, Calif.: Stanford University Press, 1999); Talal Asad, "Notes on Body Pain and Truth in Medieval Christian Ritual," *Economy and Society* 12 (1983): 287–327.

31. Martin S. Pernick, *A Calculus of Suffering: Pain, Professionalism, and Anesthesia in Nineteenth-Century America* (New York: Columbia University Press, 1985), 45; David B. Morris, *The Culture of Pain* (Berkeley: University of California Press, 1991); Lucy Bending, *The Representation of Bodily Pain in Late Nineteenth-Century English Culture* (Oxford: Oxford University Press, 2000).

32. Sharon Traweek introduces the phrase "culture of no culture" in her study of high-energy physics in Japan and the United States. See *Beamtimes and Lifetimes: The World of High Energy Physicists* (Cambridge, Mass.: Harvard University Press, 1988), 162.

33. On the rise of the "aperspectival observer" in this period, see Evelyn Fox Keller, "The Dilemma of Scientific Subjectivity in Postvital Culture," in *The Disunity of Science: Boundaries, Contexts, and Power,* ed. Peter Galison and David J. Stump (Stanford, Calif.: Stanford University Press, 1996), 419; Keller, *Reflections on Gender and Science* (New Haven, Conn.: Yale University Press, 1985); and Keller, "The Paradox of Scientific Subjectivity," *Annals of Scholarship* 9, no. 2 (1992): 135–54.

34. Michel Foucault, "On the Genealogy of Ethics: An Overview of Work in Progress," in *Ethics: Subjectivity and Truth; Essential Works of Foucault, 1954–1984,* ed. Paul Rabinow, trans. Robert Hurley et al. (New York: New Press, 1997), 279.

35. Steven Shapin's studies of early modern natural philosophy have been particularly important in revealing the continuing significance of questions of moral virtue in the ability to speak the truth. See Steven Shapin and Simon Schaffer, *Leviathan and the Air-Pump: Hobbes, Boyle, and the Experimental Life* (Princeton, N.J.: Princeton University Press, 1985); Shapin, *A Social History of Truth*; Shapin, "'The Mind Is Its Own Place': Science and Solitude in Seventeenth-Century England," *Science in Context* 4, no. 1 (1990): 191–218; and Shapin, "The Philosopher and the Chicken: On the Dietetics of Disembodied Knowledge," in *Science Incarnate: Historical Embodiments of Natural Knowledge,* ed. Steven Shapin and Christopher Lawrence (Chicago: University of Chicago Press, 1998), 21–50. Other historians of early modern science have joined Shapin in bringing new attention to the bodily practices of the knowing subject. See, for example, Jan Golinski, "The Care of the Self and the Masculine Birth of Science," *History of Science* 40 (2002): 125–45; Mario Biagioli, "Tacit Knowledge, Courtliness, and the Scientist's Body," in *Choreographing History,* ed. Susan Leigh Foster (Bloomington: Indiana University Press, 1995), 69–81; J.R.R. Christie, "The Paracelsian Body," in *Paracelsus: The Man and His Reputation, His Ideas and Their Transformation,* ed. Ole Peter Grell (Leiden: Brill, 1998), 269–91; Simon Schaffer, "Piety, Physic and Prodigious Abstinence," in *Religio Medici: Medicine and Religion in Seventeenth-Century England,* ed. Ole Peter Grell and Andrew Cunningham (Aldershot, U.K.: Scolar, 1996), 171–203; Adrian Johns, "The Physiology of Reading and the Anatomy of Enthusiasm," in *Religio Medici,* 136–70; and Stuart Walker Strickland, "The Ideology of Self-Knowledge and the Practice of Self-Experimentation," *Eighteenth-Century Studies* 31, no. 4 (1998): 453–71.

36. George Levine, for instance, suggests that Descartes did not break with preceding ascetic traditions but disguised these traditions—"the virtues of ideal self-surrender, of endurance of pain and sacrifice"—in method, so that "epistemological success becomes a morality of its own, without acknowledgement" (*Dying to Know: Scientific Epistemology and Narrative in Victorian England* [Chicago: University of Chicago Press, 2002], 55). Golinski critiques Foucault's claim more explicitly; see "The Care of the Self."

37. Daston, "Objectivity and the Scientific Self," paper delivered at Rutgers University, 24 February 2003, p. 2.

38. Ibid., 3; Lorraine Daston, "Objectivity and the Escape from Perspective," *Social Studies of Science* 22 (1992): 614; Levine, *Dying to Know,* 5.

39. I thank Andy Lakoff for sharpening this point for me.

40. See, for example, Charles-Edouard Brown-Séquard, "Des Effets produits chez l'homme par des injections sous-cutanées une liquide retiré des testicules frais de cobaye et de chien," *Comptes rendus hebdomadaires de séances et mémoires de la Sociéte de Biologie* 1

(1889): 415–19; F. A. von Humboldt, *Versuche über die gereizte Muskel- und Nervenfaser* (Berlin, 1797), cited in Simon Schaffer, "Genius," in *Romanticism and the Sciences*, ed. Andrew Cunningham and Nicholas Jardine (Cambridge, U.K.: Cambridge University Press, 1990), 92; and Gaston Tissandier, *Les martyrs de la science* (Paris: Dreyfous, 1880).

41. Quoted in S. S. Schweber, "Scientists As Intellectuals: The Early Victorians," in *Victorian Science and Victorian Values: Literary Perspectives*, ed. James Paradis and Thomas Postlewait (New Brunswick, N.J.: Rutgers University Press, 1985), 16.

42. W.H.R. Rivers and Henry Head, "A Human Experiment in Nerve Division," *Brain* 31 (November 1908): 389.

43. Henry Ward Beecher, "Suffering, the Measure of Worth," cited in Pernick, *Calculus of Suffering*, 279–80.

44. On postwar nostalgia, see David W. Blight, *Race and Reunion: The Civil War in American Memory* (Cambridge, Mass.: Harvard University Press, 2001); and Gerald F. Linderman, *Embattled Courage: The Experience of Combat in the American Civil War* (New York: Free Press, 1987), 291. On the artist, see T. S. Eliot, "Tradition and Individual Talent," in *The Sacred Wood: Essays on Poetry and Criticism* (London: Methuen, 1948 [1920]), 52–53; on the Christian, see Charles A. Briggs, "The Salvation Army," *North American Review* 159 (1894): 697–710. For discussion of the general prevalence of sacrifice in fin-de-siècle letters, see Susan L. Mizruchi, *The Science of Sacrifice: American Literature and Modern Social Theory* (Princeton, N.J.: Princeton University Press, 1998); and Alice Owen Letvin, *Sacrifice in the Surrealist Novel: The Impact of Early Theories of Primitive Religion on the Depiction of Violence in Modern Fiction* (New York: Garland, 1990).

45. John K. Noyes, *The Mastery of Submission: Inventions of Masochism* (Ithaca, N.Y.: Cornell University Press, 1997); Carol Siegel, *Male Masochism: Modern Revisions of the Story of Love* (Bloomington: Indiana University Press, 1995).

46. Letvin, *Sacrifice in the Surrealist Novel*; Mizruchi, *Science of Sacrifice.*

47. Charles Kingsley, "The Study of Physical Science: A Lecture to Young Men," *Popular Science Monthly* 1 (August 1872): 454.

48. Elaine Scarry, "The Merging of Bodies and Artifacts in the Social Contract," in *Culture on the Brink: Ideologies of Technology*, ed. Gretchen Bender and Timothy Druckrey (Seattle: Bay, 1994), 85; Scarry, "Consent and the Body: Injury, Departure, and Desire," *New Literary History* 21 (autumn 1990): 867–96.

49. Toni Morrison, *Beloved* (New York: Signet, 1991 [1987]), 172. See also Cheryl I. Harris, "Whiteness As Property," in *Critical Race Theory: The Key Writings That Formed the Movement*, ed. Kimberlé Crenshaw, Neil Gotanda, Gary Peller, and Kendall Thomas (New York: New Press, 1995).

50. William James, *The Essential Writings*, ed. Bruce W. Wilshire (Albany: State University of New York Press, 1989), 352. See also Charles W. Mills, *The Racial Contract* (Ithaca, N.Y.: Cornell University Press, 1999), 58; and Londa Schiebinger, *Has Feminism Changed Science?* (Cambridge, Mass.: Harvard University Press, 1999), 74.

51. This approach is guided by Joan W. Scott's landmark essay, "Experience," in *Feminists Theorize the Political*, ed. Judith Butler and Joan W. Scott (New York: Routledge, 1992), 22–40.

52. See, for example, Gail Bederman, *Manliness and Civilization: A Cultural History of Gender and Race in the United States, 1880–1917* (Chicago: University of Chicago Press, 1995); Mark C. Carnes and Clyde Griffen, eds., *Meanings for Manhood: Constructions of Masculinity in Victorian America* (Chicago: University of Chicago Press, 1990); Amy Kaplan and Donald E. Pease, eds., *Cultures of United States Imperialism* (Durham, N.C.: Duke University Press, 1993), 129–63; T. J. Jackson Lears, *No Place of Grace: Antimodernism and*

the *Transformation of American Culture* (New York: Pantheon, 1981); Mark Seltzer, *Bodies and Machines* (New York: Routledge, 1992); R. Marie Griffith, "Apostles of Abstinence: Fasting and Masculinity during the Progressive Era," *American Quarterly* 52 (December 2000): 599–638; Maurizia Boscagli, *Eye on the Flesh: Fashions of Masculinity in the Early Twentieth Century* (Boulder, Colo.: Westview, 1996); and Warwick Anderson, "The Trespass Speaks: White Masculinity and Colonial Breakdown," *American Historical Review* 102 (December 1997): 1343–70.

53. See, for example, Susan E. Lederer, "The Controversy over Animal Experimentation in America, 1880–1914," in *Vivisection in Historical Perspective*, ed. Nicholaas A. Rupke (London: Croom Helm, 1987), 236–58; Lederer, "Political Animals: The Shaping of Biomedical Research Literature in Twentieth-Century America," *Isis* 83 (March 1992): 61–79; Lederer, *Subjected to Science*; Jordan Goodman, Anthony McElligott, and Lara Marks, eds., *Useful Bodies: Humans in the Service of Medical Science in the Twentieth Century* (Baltimore: Johns Hopkins University Press, 2003); Helena M. Pycior, Nancy G. Slack, and Pnina G. Abir-Am, eds., *Creative Couples in the Sciences* (New Brunswick, N.J.: Rutgers University Press, 1996); Shapin, "Invisible Technicians: Masters, Servants, and the Making of Experimental Knowledge," in *Social History of Truth*, 355–407; Susan M. Reverby, ed., *Tuskegee's Truths: Rethinking the Tuskegee Syphilis Study* (Chapel Hill: University of North Carolina Press, 2000); and Andrew Goliszek, *In the Name of Science: A History of Secret Programs, Medical Research, and Human Experimentation* (New York: St. Martin's, 2003).

54. David Savran, "The Sadomasochist in the Closet: White Masculinity and the Culture of Victimization," *differences* 8 (1996): 145–46.

55. For further discussion of the importance of technologies of individual domination and their relationship to the domination of others, see Foucault, "Technologies of the Self," in *Ethics*, 225.

56. "Resolutions of the International Geodetic Commission in Relation to the Unification of Longitudes and of Time," *Science* 2 (28 December 1883): 815.

57. "Self-Sacrifice for the Sake of Science," 386.

58. "The Longing of Science to Remain Useless," *Current Opinion* 71 (July 1921): 89.

59. Francis B. Summer, "Some Perils Which Confront Us As Scientists," *Scientific Monthly* 8 (March 1919): 258–74.

60. Slosson, "Relative Value of Life and Learning," 7. Author of a number of books on popular science, Slosson became director of Science Service, a syndicate that circulated updates on science to newspapers and magazines, upon its organization in 1920.

61. Marcel Mauss is the most important analyst of the nonreciprocal exchange known as the gift; see *The Gift: The Form and Reason for Exchange in Archaic Societies*, trans. W. D. Hall (New York: Norton, 1990 [1924]). My approach is most influenced by the extensions and complications of Mauss's analysis offered by George Bataille and Jacques Derrida. See Bataille, *The Accursed Share: An Essay on General Economy*, vol. 1, trans. Robert Hurley (New York: Zone, 1991); Bataille, *Visions of Excess: Selected Writings, 1927–1939*, ed. and trans. Allan Stoekl (Minneapolis: University of Minnesota Press, 1985); and Derrida, *Given Time: I. Counterfeit Money*, trans. Peggy Kamuf (Chicago: University of Chicago, 1992), especially chap. 2, "The Madness of Economic Reason: A Gift without Present."

62. Slosson, "Relative Value of Life and Learning."

63. Rather than presuming a universal principle of self-interest, in other words, this study seeks to explore changing understandings of reasonable action. As Albert O. Hirschman reminds us, "interest" itself is a modern concept (see *The Passions and the Interests: Political Arguments for Capitalism before its Triumph* [Princeton, N.J.: Princeton

University Press, 1977]). For further discussion of the importance of actors' perspectives on action, see Michael Lynch, "Springs of Action or Vocabularies of Motive?" in *Wellsprings of Achievement: Cultural and Economic Dynamics in Early Modern England and Japan*, ed. Penelope Gouk (Aldershot, U.K.: Variorium, 1995), 94–113; C. Wright Mills, "Situated Actions and Vocabularies of Motive," in *Power, People, and Politics* (New York: Oxford University Press, 1963), 439–68; Pierre Bourdieu, *Outline of a Theory of Practice*, trans. Richard Nice (Cambridge, U.K.: Cambridge University Press, 1977); and Rebecca Herzig, "On Performance, Productivity, and Vocabularies of Motive in Recent Studies of Science," *Feminist Theory* 5, no. 2 (2004): 127–47.

64. For a particularly lucid explanation of why "normative repertoires" might be treated as "a constitutive feature of 'real' behavior," see Shapin, *Social History of Truth*, xxi. My understanding of the inseparability of words and wounds is indebted to Jill Lepore's analysis in *The Name of War: King Philip's War and the Origins of American Identity* (New York: Vintage, 1999); and to Karen Barad's discussion of "agential realism" in "Getting Real: Technoscientific Practices and the Materialization of Reality," *differences* 10, no. 2 (1998): 87–128.

65. Moreover, a single act might occupy two or more forms of reasoning successively or even at once, as John Tresch helped me to understand.

66. Stephen J. Collier and Andrew Lakoff, "On Regimes of Living," in *Global Assemblages: Technology, Politics, and Ethics As Anthropological Problems*, ed. Aihwa Ong and Stephen J. Collier (New York: Blackwell, 2005), 22. See also Foucault, "On the Genealogy of Ethics," in *Ethics*; Foucault, *The Use of Pleasure*, vol. 2 of *The History of Sexuality*, trans. Robert Hurley (New York: Vintage, 1990), especially 26–30; Paul Rabinow, "Science As a Practice: Ethos, Logos, Pathos," in *Essays on the Anthropology of Reason* (Princeton, N.J.: Princeton University Press, 1996), 3–27; and Rabinow, *Anthropos Today: Reflections on Modern Equipment* (Princeton, N.J.: Princeton University Press, 2003), especially 101–2. On suffering as a problem of ethics, see Emmanuel Levinas, "Useless Suffering," in *The Provocation of Levinas: Rethinking the Other*, ed. Robert Bernasconi and David Wood (London: Routledge, 1988), 156–67.

67. Max Weber, "Science As a Vocation," in *From Max Weber: Essays in Sociology*, trans. and ed. H. H. Gerth and C. Wright Mills (New York: Oxford University Press, 1946), 143.

68. Emile Durkheim, for example, argued that faith in science "does not differ essentially from religious faith." "In the last resort," he wrote, "the value which we attribute to science depends upon the idea which we collectively form of its nature and role in life. . . . science continues to be dependent upon opinion at the very moment when it seems to be making its laws; for, as we have already shown, it is from opinion that it holds the force necessary to act upon opinion." See *The Elementary Forms of Religious Life*, trans. Joseph Ward Swain (London: Allen and Unwin, 1915), 438.

69. William James, "The Dilemma of Determinism," in *The Will to Believe and Other Essays in Popular Philosophy* (New York: Longmans, Green, 1919), 147.

70. Friedrich Nietzsche, *The Gay Science*, trans. Walter Kaufmann (New York: Vintage, 1974 [1882]), 281. Also see Nietzsche, *On the Genealogy of Morals*, trans. Walter Kaufmann and P. J. Hollingdale (New York: Vintage, 1967 [1887]).

CHAPTER I   WILLING CAPTIVES

1. Charles K. Mills, *Mental Over-Work and Premature Disease among Public and Professional Men* (Washington, D.C.: Smithsonian, 1885), 14. I am grateful to Lorraine Daston for bringing this text to my attention.

2. G. Stanley Hall, "Overpressure in Schools," *Nation* 41 (22 October 1885): 338.

3. Mary Poovey, "Speaking of the Body: Mid-Victorian Constructions of Female Desire," in *Body/Politics: Women and the Discourses of Science,* ed. Mary Jacobus, Evelyn Fox Keller, and Sally Shuttlesworth (New York: Routledge, 1990), especially 33; Judith Rowbotham, " 'Soldiers of Christ?' Images of Female Missionaries in Late Nineteenth-Century Britain: Issues of Heroism and Martyrdom," *Gender and History* 12 (April 2000): 82–106; Drew Gilpin Faust, "Altars of Sacrifice: Confederate Women and the Narratives of War," *Journal of American History* 76 (March 1990): 1200–28; Pernick, *Calculus of Suffering,* 47–49.

4. My use of *sex* (rather than the more common term *gender*) may benefit from a bit of explanation. A number of critics have pointed out that the term *gender,* adopted by 1970s feminists as a way to draw attention to the arbitrariness of perceived differences between the sexes, often has the unfortunate effect of naturalizing the bodies on which cultural differences are allegedly based. Once we examine the presumption of sexual dimorphism itself, we learn, as Judith Butler has put it, that "perhaps this construct called 'sex' is as culturally constructed as gender; indeed, perhaps it was always already gender, with the consequence that the distinction between sex and gender turns out to be no distinction at all." (See Butler, *Gender Trouble: Feminism and the Subversion of Identity* [New York: Routledge, 1990]), 7.) Rather than employing the term *gender* in this study (and further confusing the questions of human nature at the heart of these debates), I therefore use *sex* in accordance with period sources. Throughout, I attempt to show how particular practices helped maintain the appearance of innate sexual difference. For further discussion of the category of gender, particularly in relation to contested definitions of the human, see Judith Butler, *Undoing Gender* (New York: Routledge, 2004).

5. In this study I use the phrases *self-ownership, self-possession,* and *property-in-the-person* more or less interchangeably. Carole Pateman has cautioned against such elision of terms, noting that recent discussions of self-ownership tend to efface the divisive political questions contained in the concept of property-in-the-person employed by Locke and later critics. I here endeavor to keep the political implications of proprietal selfhood at the center of my analysis, while also using the most readable language possible. See Pateman, "Self-Ownership and Property in the Person: Democratization and a Tale of Two Concepts," *Journal of Political Philosophy* 10, no. 1 (2002): 20–53. I also interchange the terms *willing* and *voluntary* here, in accordance with my nineteenth-century sources. Philosophers have distinguished the meanings of these and related concepts; see, for example, Thomas Pink and M.W.F. Stone, eds., *The Will and Human Action: From Antiquity to the Present* (London: Routledge, 2004).

6. John Foxe, *Foxe's Book of Martyrs* (Springdale, Pa.: Whitaker, 1981 [1563]); Cotton Mather, *The Sacrificer: An Essay Upon the Sacrifices Wherewith a Christian, laying a Claim to an Holy Priesthood, Endeavours to Glorify God* (Boston: Fleet, 1714); Mather, *A Brief Essay Upon the Cross* (Boston: Allen, 1714); Sydney E. Ahlstrom, *A Religious History of the American People* (New Haven, Conn.: Yale University Press, 1972); Hazel Catherine Wolf, *On Freedom's Altar: The Martyr Complex in the Abolition Movement* (Madison: University of Wisconsin Press, 1952).

7. Perry Miller, "The Marrow of Puritan Divinity," in *Errand into the Wilderness* (Cambridge, Mass.: Harvard University Press, 1956), 62. Also see Daniel J. Elazar, ed., *Covenant in the Nineteenth Century: The Decline of an American Tradition* (Lanham, Md.: Rowman and Littlefield, 1994). I thank Susan Conrad for reminding me of the importance of covenant in establishing the conditions for sacrifice.

8. Miller, *Errand*, 63.

9. Cited in Foner, "Meaning of Freedom," 438. Also see William Warren Sweet, *The Story of Religions in America* (New York: Harper, 1930), 265.

10. Jonathan Edwards, *The Works of President Edwards* (New York: Leavitt and Allen, 1851), 4:370.

11. Dan McKanan, *Identifying the Image of God: Radical Christians and Nonviolent Power in the Antebellum United States* (New York: Oxford University Press, 2002); Timothy L. Smith, *Revivalism and Social Reform: American Protestantism on the Eve of the Civil War* (New York: Harper and Row, 1957); Cynthia Eagle Russett, *Darwin in America: The Intellectual Response, 1865–1912* (San Francisco: Freeman, 1976); Sweet, *Story of Religions*; Ahlstrom, *Religious History*.

12. Adams to James Sullivan, 26 May 1776, cited in Linda K. Kerber, "'Ourselves and Our Daughters Forever': Women and the Constitution, 1787–1876," in *One Woman, One Vote: Rediscovering the Woman Suffrage Movement*, ed. Marjorie Spruill Wheeler (Troutdale, Or.: NewSage, 1995), 25

13. As Foner notes, the effect of this change was to supplant class with race "as the boundary defining which American men were to enjoy political freedom." See "Meaning of Freedom," 442, 443.

14. John Locke, "On Property," in *Two Treatises of Government*, ed. Peter Laslett (Cambridge, U.K.: Cambridge University Press, 1988), 287.

15. Georg Simmel, "A Chapter in the Philosophy of Value," *American Journal of Sociology* 5 (March 1900): 582. For a fuller discussion of contract with nonhuman agents, see Michel Serres, *The Natural Contract* (Ann Arbor: University of Michigan Press, 1995), especially 27–50.

16. Locke, *Two Treatises*, 284.

17. Alan Hyde, *Bodies of Law* (Princeton, N.J.: Princeton University Press, 1997); Radhika Rao, "Property, Privacy, and the Human Body," *Boston University Law Review* 80 (April 2000): 359–460; Margaret Jean Radin, *Reinterpreting Property* (Chicago: University of Chicago Press, 1993), chap. 1.

18. G.W.F. Hegel, "The Encyclopaedia of Philosophical Sciences, Part I," in *The Encyclopaedia Logic*, trans. Theodore F. Geraets, W.A. Suchting, and H. S. Harris (Indianapolis: Hackett, 1991), 57.

19. Miller, *Errand*, 168.

20. Carole Pateman, *The Sexual Contract* (Stanford, Calif.: Stanford University Press, 1988), 55–56.

21. Ibid., 15.

22. C. B. Macpherson, *The Political Theory of Possessive Individualism: Hobbes to Locke* (Oxford: Clarendon, 1962), 264. Building on the promotion of "selfe propriety" evident among seventeenth-century Leveller radicals, Rosalind Pollack Petchesky offers a provocative alternate reading of the political possibilities of proprietal selfhood in "The Body As Property: A Feminist Re-vision," in *Conceiving the New World Order: The Global Politics of Reproduction*, ed. Faye D. Ginsberg and Rayna Rapp (Berkeley: University of California Press, 1995), 387–406.

23. Mills, *Racial Contract*, 3. See also Pateman, *Sexual Contract*; Pateman, "Self-Ownership and Property"; Iris Marion Young, "Impartiality and the Civic Public: Some Implications of Feminist Critique of Moral and Political Theory," in *Feminism As Critique*, ed. Seyla Benhabib and Drucilla Cornell (Minneapolis: University of Minnesota, 1987), 57–76; Charles W. Mills, "Black Trash," in *Faces of Environmental Racism*, ed. Laura Westra and Bill E. Lawson (Boulder, Colo.: Rowman and Littlefield, 2001), 73–91; and Petchesky, "Body As Property."

24. Cited in James Oakes, *Slavery and Freedom: An Interpretation of the Old South* (New York: Knopf, 1990), 80.

25. Kerber, "Ourselves and Our Daughters Forever," 21–36; People v. Coffman, 184 Calif. 230 (1864); Myra C. Glenn, *Campaigns against Corporal Punishment: Prisoners, Sailors, Women, and Children in Antebellum America* (Albany: State University of New York Press, 1984).

26. *United States v. Cargo of the Brig Malek Adhel*, 43 U.S. (2 How.) 210 (1844). See also Christopher D. Stone, *Should Trees Have Standing? Toward Legal Rights for Natural Objects* (Palo Alto, Calif.: Tioga, 1988).

27. Hortense J. Spillers, "Mama's Baby, Papa's Maybe: An American Grammar Book," *diacritics* 17 (summer 1987): 67.

28. Ibid., 67. The point here is not that whites were allotted selfhood while blacks were not. Whiteness and blackness were no more naturally given conditions than were historical arrangements of bondage. The existence of so-called black slaves with fairer skin than their putatively white owners points to the contingencies of racial embodiment: "white" is not a preordained state but an effect of social relations. What matters in this context is that whiteness was placed at the center of definitions of property and hence at the center of definitions of freedom. As legal theorist Cheryl I. Harris argues, since persons defined as white could not be owned by another, whiteness itself became the central sign of liberal selfhood. Once "slavery was conflated with racial identity," whiteness became "the characteristic, the attribute, the property of free human beings." See Harris, "Whiteness As Property," in *Critical Race Theory: The Key Writings That Formed the Movement*, ed. Kimberlé Crenshaw, Neil Gotanda, Gary Peller, and Kendall Thomas (New York: New Press, 1995), 279. On the materialization of whiteness, also see Walter Johnson, "The Slave Trader, the White Slave, and the Politics of Racial Determination in the 1850s," *Journal of American History* 87 (June 2000): 13–38; and Jonathan Xavier Inda, "Performativity, Materiality, and the Racial Body," *Latino Studies Journal* 11 (fall 2000): 74–99.

29. N. Katherine Hayles, *How We Became Posthuman: Virtual Bodies in Cybernetics, Literature, and Informatics* (Chicago: University of Chicago Press, 1999), 4. For a fuller discussion of the relationships among sex, race, corporeality, and political identity in this period, see Karen Sánchez-Eppler, *Touching Liberty: Abolition, Feminism, and the Politics of the Body* (Berkeley: University of California Press, 1993).

30. Hartmann, *Scenes of Subjection*, 36ff.

31. A.P. Merrill, "An Essay on Some of the Distinctive Peculiarities of the Negro Race," *Memphis Medical Recorder* 4 (1855), cited in Pernick, *Calculus of Suffering*, 155. As historian Todd Savitt points out, claims that Negroes did not feel pain as whites did hardly prohibited physicians from applying experimental findings drawn from slave studies to their white patients. See Savitt, "The Use of Blacks for Medical Experimentation and Demonstration in the Old South," *Journal of Southern History* 48 (August 1982): 332.

32. Cited in Pernick, *Calculus of Suffering*, 156.

33. Esther Cohen, "The Animated Pain of the Body," *American Historical Review* 105 (February 2000): 38; Daniel de Moulin, "A Historical-Phenomenological Study of Bodily Pain in Western Man," *Bulletin of the History of Medicine* 48 (1984): 540–70; Elizabeth B. Clark, "'The Sacred Rights of the Weak': Pain, Sympathy, and the Origins of Humanitarian Sensibility," *Journal of American History* 82, no. 2 (1995): 474; Pernick, *Calculus of Suffering*, 155–56; Diane Price Herndl, "The Invisible (Invalid) Woman: African-American Women, Illness, and Nineteenth-Century Narrative," in *Women and Health in America*, ed. Judith Walker Leavitt, 2d ed. (Madison: University of Wisconsin Press, 1999),

131–45; Savitt, "Use of Blacks," 331–48; James Denny Guillory, "The Pro-Slavery Arguments of Dr. Samuel A. Cartwright," *Louisiana History* 9 (1968): 209–27; Amy Dru Stanley, "'The Right to Possess All the Faculties That God Has Given': Possessive Individualism, Slave Women, and Abolitionist Thought," in *Moral Problems in American Life: New Perspectives on Cultural History*, ed. Karen Halttunen and Lewis Perry (Ithaca, N.Y.: Cornell University Press, 1998), 123–43.

34. Spillers, "Mama's Baby," 67. See also Walter Johnson, *Soul by Soul: Life Inside the Antebellum Slave Market* (Cambridge, Mass.: Harvard University Press, 1999).

35. Cited in Harris, "Whiteness As Property," 279.

36. People v. George W. Hall, 4 Calif. 399 (1854).

37. Charles V. Carnegie, "The Dundus and the Nation," *Cultural Anthropology* 11, no. 4 (1996): 481.

38. Donna J. Haraway, *Modest_Witness@Second_Millennium.FemaleMan©_Meets_OncoMouse™: Feminism and Technoscience* (New York: Routledge, 1997), 38.

39. Derrida, *Given Time*, 37–42.

40. George M. Frederickson, *Inner Civil War: Northern Intellectuals and the Crisis of the Union* (New York: Harper and Row, 1965), 81; Ernst H. Kantorowicz, *The King's Two Bodies: A Study in Mediaeval Political Theology* (Princeton, N.J.: Princeton University Press, 1997 [1957]).

41. Frank Moore, *Women of the War: Their Heroism and Self-Sacrifice* (Hartford, Conn.: Scranton, 1866). Also see Faust, "Altars of Sacrifice"; Frederickson, *Inner Civil War*; James M. McPherson, *For Cause and Comrades: Why Men Fought in the Civil War* (New York: Oxford University Press, 1997); Wolf, *On Freedom's Altar*; Gary Laderman, *The Sacred Remains: American Attitudes Toward Death, 1799–1883* (New Haven, Conn.: Yale University Press, 1993), especially chap. 10; and Elizabeth D. Leonard, *Yankee Women: Gender Battles in the Civil War* (New York: Norton, 1994), especially 170–72.

42. *Record of News, History and Literature*, 16 July 1863, p. 37, cited in Faust, "Altars of Sacrifice," 1213.

43. Gail Hamilton, "A Call to My Country-Women," *Atlantic Monthly* 11 (March 1863): 346. Frederickson reports that Hamilton was the pen name for Mary Abigail Dodge (see *Inner Civil War*, 248, note 11).

44. Norton to G. W. Curtis, 11 May 1862, quoted in Kermit Vanderbilt, *Charles Eliot Norton: Apostle of Culture in a Democracy* (Cambridge, Mass.: Belknap, 1959), 81; Faust, "Altars of Sacrifice"; Wolf, *On Freedom's Altar*. On the larger paradoxes of humanitarian attitudes toward pain in the antebellum period, see Karen Halttunen, "Humanitarianism and the Pornography of Pain in Anglo-American Culture," *American Historical Review* 100 (April 1995): 303–34.

45. Frederickson, *Inner Civil War*, 82.

46. Ralph Waldo Emerson, *Poems by Ralph Waldo Emerson*, in *The Complete Works of Ralph Waldo Emerson* (Boston: Houghton Mifflin, 1904), 9:207.

47. Jeffrey P. Hantover, "The Boy Scouts and the Validation of Masculinity," in *The American Man*, ed. Elizabeth H. Pleck and Joseph H. Pleck (Englewood Cliffs, N.J.: Prentice Hall, 1980), 290; Register, "Everyday Peter Pans," 203; Bederman, *Manliness and Civilization*, 12; David R. Roediger, *The Wages of Whiteness: Race and the Making of the American Working Class* (London: Verso, 1991), 96; Herbert G. Gutman, *Work, Culture, and Society in Industrializing America* (New York: Knopf, 1977), 3–78; Stanley, *From Bondage to Contract*; Angela Y. Davis, "From the Prison of Slavery to the Slavery of Prison: Frederick Douglass and the Convict Lease System," in *Frederick Douglass: A Critical Reader*, ed. Bill E. Lawson and Frank M. Kirkland (Malden, Mass.: Blackwell, 1999), 339–62.

48. William Graham Sumner, *What Social Classes Owe to Each Other* (New York: Harper, 1883), 65.

49. Hartmann, *Scenes of Subjection*, 125, 140. See also Robyn Wiegman, "Intimate Publics: Race, Property, and Personhood," *American Literature* 74 (December 2002): 859–85.

50. Reva B. Siegal, "Home As Work: The First Women's Rights Claims Concerning Wives' Household Labor, 1850–1880," *Yale Law Journal* 103 (March 1994): 1073–1217; Amy Dru Stanley, "Conjugal Bonds and Wage Labor: Rights of Contract in the Age of Emancipation," *Journal of American History* 75 (September 1988): 471–500; Foner, "Meaning of Freedom"; Kerber, "Ourselves and Our Daughters Forever," 30–31; Stanley, *From Bondage to Contract*, 199.

51. Cited in Foner, "Meaning of Freedom," 455–56.

52. The spate of naturalization cases after 1878 can be said to reflect the increasing importance of national citizenship in the last quarter of the nineteenth century. See Ian Haney López, "Racial Restrictions in the Law of Citizenship" and "The Prerequisite Cases," in *Racial Classification and History*, ed. E. Nathaniel Gates (New York: Garland, 1997), 109–60.

53. Kerber, "Ourselves and Our Daughters Forever," 31–32.

54. Stuart Creighton Miller, *The Unwelcome Immigrant: The American Image of the Chinese, 1785–1882* (Berkeley: University of California Press, 1969), 152–54.

55. Reported in Thomas J. Vivian, "John Chinaman in San Francisco," *Scribner's Monthly* 12 (1876): 863.

56. Miller, *Unwelcome Immigrant*, 154. Also see Stanley, *From Bondage to Contract*, 218–63.

57. The editor of the American edition of Count Arthur de Gobineau's influential 1856 exposition on race found the role of sex so crucial to an understanding of evolutionary standing that he added his own seven-page treatment of the subject to the count's essay. (See H. Hotz, "Note to the Preceding Chapter," in de Gobineau, *The Moral and Intellectual Diversity of the Races* [Philadelphia: Lippincott, 1856], 384–90.) As historian Emily S. Rosenberg shows, the status of women is often invoked in treatises on racial difference, summoned to legitimate humanitarian projects of rescue. For a fuller discussion, see Rosenberg, "Rescuing Women and Children," *Journal of American History* 89 (September 2002): 456–65.

58. Cited in Eithne Luibhéid, *Entry Denied: Controlling Sexuality at the Border* (Minneapolis: University of Minnesota Press, 2002), 41–42.

59. Santa Clara County v. Southern Pacific Railroad Company, 118 U.S. 394 (1886); David Korten, *The Post-Corporate World: Life after Capitalism* (San Francisco: Berrett-Koehler and Kumarian, 1999), 185–86.

60. Charles Eliot Norton, "The Advantages of Defeat," *Atlantic Monthly* 8 (September 1861): 364.

61. Georges Clemenceau, *American Reconstruction, 1865–1870*, ed. Fernand Baldensperger, trans. Margaret MacVeagh (New York: Da Capo, 1969), 297–98. Also see Russett, *Darwin in America*.

62. Lewis Henry Morgan, *Ancient Society; or, Researches in the Lines of Human Progress from Savagery through Barbarism to Civilization*, ed. Eleanor Burke Leacock (Gloucester, Mass.: Smith, 1974 [1877]), 563.

63. On the interpenetration of race, gender, age, and class in the rhetoric of civilization, see Bederman, *Manliness and Civilization*; George W. Stocking, Jr., *Race, Culture, and Evolution* (New York: Free Press, 1968); and Thomas C. Patterson, *Inventing Western Civilization* (New York: Monthly Review Press, 1997).

64. Charles Darwin, *The Expression of the Emotions in Man and Animals* (New York: Green-

wood, 1955 [1872]), 356; E. T. Seton, *Boy Scouts of America: A Handbook of Woodcraft, Scouting and Lifecraft* (New York: Doubleday, Page, 1910); Eugen Sandow, *Body-Building, or Man in the Making: How to Become Healthy and Strong* (London: Gale and Polden, 1904); Griffith, "Apostles of Abstinence"; John Kasson, *Rudeness and Civility: Manners in Nineteenth-Century Urban America* (New York: Hill and Wang, 1990), especially chap. 5.

65. Sigmund Freud, "On the Universal Tendency to Debasement in the Sphere of Love," in *The Standard Edition to the Complete Works of Sigmund Freud*, trans. James Strachey (London: Hogarth, 1957 [1912]), 11:190.

66. Henry M. Lyman, *Artificial Anaesthesia and Anaesthetics* (New York: Wood, 1881), 68; Joseph Taber Johnson, "On Some of the Apparent Peculiarities of Parturition in the Negro Race, with Remarks on Race Pelves in General," *American Journal of Obstetrics* 8 (January 1875): 88–123; Laura Briggs, "The Race of Hysteria: 'Overcivilization' and the 'Savage' Woman in Late Nineteenth-Century Obstetrics and Gynecology," *American Quarterly* 52 (June 2000): 246–73.

67. Samuel L. Brickner, "On the Physiological Character of the Pain of Parturition," *Gaillard's Medical Journal* 72 (January 1900): 795.

68. Henry Jacob Bigelow, *Surgical Anaesthesia: Addresses and Other Papers* (Boston: Little, Brown, 1900), 374.

69. Pernick, *Calculus of Suffering*, 157, 318; Nancy Stepan, *The Idea of Race in Science: Great Britain, 1800–1960* (Hamden, Conn.: Archon, 1982).

70. S. Weir Mitchell, "Civilization and Pain," *Annals of Hygiene* 7, no. 1 (1892): 26, cited in Morris, *Culture of Pain*, 39.

71. James Turner, *Reckoning with the Beast: Animals, Pain, and Humanity in the Victorian Mind* (Baltimore: Johns Hopkins University Press, 1980); Miller, *Unwelcome Immigrant*; Kaplan and Pease, *Cultures of United States Imperialism*.

72. E. L. Youmans, "The Higher Education of Women" (April 1874), cited in Louise Newman, ed., *Men's Ideas/Women's Realities: Popular Science, 1870–1915* (New York: Pergamon, 1985), 71.

73. Thomas H. Huxley, "Emancipation Black and White," in *Science and Education: Essays* (New York: Appleton, 1894 [1865]), 68.

74. David H. Cochran, testimony in *Vivisection: Hearing before the Senate Committee, 1900*, cited in Lederer, *Subjected to Science*, 42.

75. Lauren Berlant, "National Brands/National Body: Imitation of Life," in *Comparative American Identities: Race, Sex, and Nationality in the Modern Text*, ed. Hortense J. Spillers (New York: Routledge, 1991), 133.

CHAPTER 2   THE BONDS OF SCIENCE

1. Although I forgo incessant emphasis marks around the word, I hope it will remain evident that my repeated references to *science* as a unified and autonomous force are in accordance with period sources. Rather than reinforce the image of science as a timeless agent with its own intrinsic demands, I want to illuminate how this consequential abstraction came into being. In this regard, I am indebted to recent scholarship that foregrounds the heterogeneity of science and its practices. See, for example, Peter Galison and David J. Stump, eds., *The Disunity of Science: Boundaries, Contexts, and Power* (Stanford, Calif.: Stanford University Press, 1996); and John Law and Annemarie Mol, eds., *Complexities: Social Studies of Knowledge Practices* (Durham, N.C.: Duke University Press, 2002).

2. Mauss, *Gift*, 65.

3. Slosson, "Relative Value of Life and Learning," 7. All quotations from Slosson are from this text.

4. Derrida, "The Madness of Economic Reason: A Gift without Present," in *Given Time*, 37–42.

5. Mauss, *Gift*, 14.

6. Ibid., 18. For insightful applications of Mauss's central themes to analyses of contemporary science, see Warren O. Hagstrom, "Gift Giving As an Organizing Principle in Science," in *Science in Context: Readings in the Sociology of Science*, ed. Barry Barnes and David Edge (Cambridge, Mass.: MIT Press, 1982), 21–34; and Davis Baird, *Thing Knowledge: A Philosophy of Scientific Instruments* (Berkeley: University of California Press, 2004).

7. Derrida, *Given Time*, 37.

8. Mauss, *Gift*, 37, 42; Bataille, *Accursed Share* and *Visions of Excess*; Brown, "Dionysus in 1990," in *Apocalypse*. Ted Metcalfe drew my attention to Bataille's assignation of "paternity"; I am grateful for his vital contribution to my understanding of this point.

9. Charles Taylor, "The Person," in *The Category of the Person: Anthropology, Philosophy, History*, ed. Michael Carrithers, Steven Collins, and Steven Lukes (Cambridge, U.K.: Cambridge University Press, 1985), especially 258–61.

10. Mauss, *Gift*, 17.

11. Tore Frängsmyr, ed., *Solomon's House Revisited: The Organization and Institutionalization of Science* (Canton, Mass.: Science History Publications, 1990); David C. Lindberg and Ronald L. Numbers, eds., *The Cambridge History of Science* (Cambridge, U.K.: Cambridge University Press, 2003); Lorraine Daston and Peter Galison, "The Image of Objectivity," *Representations* 40 (1992): 81–128.

12. Steven Dick, "Simon Newcomb, William Harkness, and the Nineteenth-Century American Transit of Venus Expedition," *Journal for the History of Astronomy* 29, no. 3 (1998): 221–55; F.W.G. Baker, "The First International Polar Year, 1882–83," *Polar Record* 21 (1982): 275–85.

13. Clark A. Elliott, *History of Science in the United States: A Chronology and Research Guide* (New York: Garland, 1996), 108; Richard Hofstader and C. DeWitt Hardy, *The Development and Scope of Higher Education* (New York: Columbia University Press, 1952), 64.

14. Canonical assessments of this reorganization include Charles E. Rosenberg, *No Other Gods: On Science and American Social Thought* (Baltimore: Johns Hopkins University Press, 1997 [1976]); George H. Daniels, ed., *Nineteenth-Century American Science: A Reappraisal* (Evanston, Ill.: Northwestern University Press, 1972); Margaret W. Rossiter, *Women Scientists in America: Struggles and Strategies to 1940* (Baltimore: Johns Hopkins University Press, 1982); and Alexandra Oleson and John Voss, eds., *The Organization of Knowledge in Modern America, 1860–1920* (Baltimore: Johns Hopkins University Press, 1979).

15. Oleson and Voss, *Organization of Knowledge*, xi.

16. Bruce, *Launching of Modern American Science*, 220; "To the Senate and House of Representatives of the Commonwealth of Pennsylvania" (17 January 1800), in *Early Proceedings of the American Philosophical Society . . . 1744 to 1838* (Philadelphia: McCalla and Stavely, 1884), 292.

17. Predictably, the word's meanings varied by national and disciplinary context; I here attend only to American usage. For further discussion of changes in *science*, see Andrew Cunningham, "Getting the Game Right: Some Plain Words on the Identity and Invention of Science," *Studies in the History and Philosophy of Science* 19 (1988): 365–89; Stanley M. Guralnick, *Science and the Ante-Bellum American College* (Philadelphia: American Philosophical Society, 1975), 60; Roslynn D. Haynes, *From Faust to Strangelove: Rep-*

resentations of the Scientist in Western Literature (Baltimore: Johns Hopkins University Press, 1994), 6–8; Leonard N. Neufeldt, "The Science of Power: Emerson's Views on Science and Technology in America," *Journal of the History of Ideas* 38 (April–June 1977): 333; and Sydney Ross, "*Scientist*: The Story of a Word," *Annals of Science* 18 (June 1962): 65–85. On the history of the related word *technology*, see Leo Marx, "The Idea of 'Technology' and Postmodern Pessimism," in *Does Technology Drive History? The Dilemma of Technological Determinism*, ed. Merritt Roe Smith and Leo Marx (Cambridge, Mass.: MIT Press, 1994), 237–57.

18. "Editor's Table: Purpose and Plan of Our Enterprise," *Popular Science Monthly* 1 (May 1872): 113. See also Kingsley, "Study of Physical Science," 456.

19. T. C. Mendenhall, "The Relations of Men of Science to the General Public," *Proceedings of the American Association for the Advancement of Science* 39 (1890): 15.

20. Political philosopher Jean Bethke Elshtain points out that discussions of national sacrifice tend to blend religious, sexual, and martial metaphors freely. Subsequent chapters show that scientific self-sacrifice displays the same tendency. See Elshtain, "Sovereignty, Identity, Sacrifice," in *Reimagining the Nation*, edited by Marjorie Ringrose and Adam J. Lerner (Buckingham, U.K.: Open University Press, 1993), 161.

21. Benedict Anderson, *Imagined Communities: Reflections on the Origin and Spread of Nationalism* (London: Verso, 1991).

22. Cunningham, "Getting the Game Right," 384.

23. Daston, "Objectivity and the Escape from Perspective," 613.

24. Anderson, *Imagined Communities*, 6–7. Anderson takes up the analogous status of the nation and the person; each acquires a narrative of identity, and each might be represented as a single body (for example, Lady Liberty, Uncle Sam). Yet as Anderson recounts, there is a central difference in the narratives of person and nation, a central difference between the person and science. In secular biographies, the person has an evident beginning and end. Nations, in contrast, "have no clearly identifiable births, and their deaths, if they ever happen, are never natural" (205). Similarly, to the extent that science was personified with an independent identity, this identity was not equal to that of the finite person. Instead, like the nation, its timeline was said to be infinite.

25. Weber, "Science as a Vocation," in *From Max Weber*, 137–38.

26. Cited in Simon Schaffer, "Astronomers Mark Time," *Science in Context* 2, no. 1 (1988): 127. Focusing his account on Greenwich and other major observatories, Schaffer's periodization differs slightly from my own. The difference stems from the relatively early organization of astronomy and the effect of the war on the development of scientific institutions in the United States.

27. Cited in Daston, "Objectivity and the Escape from Perspective," 609. Also see Daston, "Objectivity and the Scientific Self"; Daston and Galison, "Image of Objectivity"; Peter Galison, "Judgement Against Objectivity," in *Picturing Science, Producing Art*, ed. Peter Galison and Caroline Jones (New York: Routledge, 1998); Levine, *Dying to Know*; Amanda Anderson, *The Powers of Distance: Cosmopolitanism and the Cultivation of Detachment* (Princeton, N.J.: Princeton University Press, 2001); Schaffer, "Astronomers Mark Time"; and Naomi Oreskes, "Objectivity of Heroism? On the Invisibility of Women in Science," *Osiris* 11 (1996): 87–113.

28. "Our Great Debt to Science," *Scientific American* 77 (4 October 1890): 217.

29. G. Stanley Hall, "Confessions of a Psychologist, Part I," *Pedagogical Seminary* 8 (1901): 119–20.

30. [William Whewell], Review of *On the Connexion of the Physical Sciences*, *Quarterly Review* 51 (March and June 1834): 59.

31. As Ross explains in *"Scientist,"* p. 75, the older phrase *man of science* persisted in Britain through 1910. When, in the 1870s, *scientist* came into wider use in the United States, some European observers derided the term as a "barbarous" Americanism, an off-shoot of a distinctly national (and lesser) style of investigative practice.

32. Simon Newcomb, "The Evolution of the Scientific Investigator," *Science* 20 (23 September 1904): 394, 393, 395.

33. Stanley, *From Bondage to Contract*, 264.

CHAPTER 3   PURISTS

1. Josiah P. Cooke, Jr., "Scientific Culture," *Popular Science Monthly* 7 (1875): 527, 525.

2. Charles Gross, "Address," in *Williams College: Centennial Anniversary, 1793–1893* (Cambridge, Mass.: Wilson, 1894), 173. On the place of forests in discourses of enlightenment, see Robert Pogue Harrison, *Forests: The Shadow of Civilization* (Chicago: University of Chicago Press, 1992), especially chap. 3.

3. G. Stanley Hall, "The University Idea," *Pedagogical Seminary* 15 (1908): 104.

4. Alexis de Tocqueville, *Democracy in America*, trans. Henry Reeve (New York: Vintage Books, 1945), 2:44.

5. See, for example, F. W. Clarke, "American Colleges *versus* American Science," *Popular Science Monthly* 9 (1876): 467–79; "Future of American Science," 1; and "Comment and Criticism," *Science* 3 (2 May 1884): 530.

6. Laurence R. Veysey, *The Emergence of the American University* (Chicago: University of Chicago Press, 1965), 122.

7. Ibid.; David A. Hollinger, "Inquiry and Uplift: Late Nineteenth-Century American Academics and the Moral Efficacy of Scientific Practice," in *The Authority of Experts*, ed. Thomas L. Haskell (Bloomington: Indiana University Press, 1984), 142–56; Julie A. Reuben, *The Making of the Modern University: Intellectual Transformation and the Marginalization of Morality* (Chicago: University of Chicago Press, 1996); Bruce A. Kimball, *The "True Professional Ideal" in America: A History* (Cambridge, Mass.: Blackwell, 1992), especially chap 4; Rosenberg, *No Other Gods*, 135–52.

8. Charles Baskerville, "Science and the People," *Science* 20 (26 August 1904): 270.

9. Perry Miller and Thomas H. Johnson, *The Puritans* (New York: Harper and Row, 1963 [1938]), 1:6.

10. Henry F. May, *Protestant Churches and Industrial America* (New York: Octagon, 1977 [1949]), xii.

11. G. Stanley Hall, *Life and Confessions of a Psychologist* (New York: Appleton, 1923), 30.

12. Robert V. Bruce, "A Statistical Profile of American Scientists, 1846–1876," in *Nineteenth-Century American Science: A Reappraisal*, ed. George H. Daniels (Evanston, Ill.: Northwestern University Press, 1972), 82. These data are somewhat hazy on the definition of *scientist*. Bruce overrides the *Dictionary of American Biography*'s classification system but does not articulate his own method for determining which nineteenth-century workers might count as scientists. On training for the ministry, also see Clark A. Elliott, "Scientists, Nature and the Self: Reading Autobiographical Writings from 19th Century America," paper presented at the History of Science Society's annual meeting, San Diego, 1997, p. 7.

13. Rosenberg, *No Other Gods*, 3. Following Max Weber's seminal argument that the Reformation endowed everyday activity with a religious significance, several generations of scholars have scrutinized the relations between Protestant fascination with individual salvation and the development of modern science. I depart from these analyses not

only in my choice of subject matter (late nineteenth-century America rather than seventeenth-century England) but also in my effort to give the practices of asceticism, particularly the degradation of the body, as much weight in the rise of science as the Calvinist doctrine of salvation. Furthermore, I suggest that the particularly precarious social order of late nineteenth-century America promoted heightened concern about collective, not merely individual, ruin. For vital discussion of these themes, see Robert K. Merton, *Science, Technology and Society in Seventeenth-Century England* (New York: Fertig, 1970); I. Bernard Cohen, ed., *Puritanism and the Rise of Modern Science: The Merton Thesis* (New Brunswick, N.J.: Rutgers University Press, 1990); Lewis S. Feuer, *The Scientific Intellectual: The Psychological and Sociological Origins of Modern Science* (New York: Basic Books, 1963); Arthur Koestler, *The Sleepwalkers: A History of Man's Changing Vision of the Universe* (New York: Macmillan, 1959); and, of course, Max Weber, *The Protestant Ethic and the Spirit of Capitalism*, trans. Talcott Parsons (New York: Scribner, 1958), 80, 62.

14. Arthur B. Stout, *Chinese Immigration and the Physiological Causes of the Decay of a Nation* (San Francisco: Agnew and Deffelbach, 1862), 8.

15. Miller, *Unwelcome Immigrant*; Luibhéid, "A Blueprint for Exclusion: The Page Law, Prostitution, and Discrimination against Chinese Women," in *Entry Denied*, 31–54; Nayan Shah, *Contagious Divides: Epidemics and Race in San Francisco's Chinatown* (Berkeley: University of California Press, 2001); K. Scott Wong, "Cultural Defenders and Brokers: Chinese Responses to the Anti-Chinese Movement," in *Claiming America: Constructing Chinese American Identities during the Exclusion Era*, ed. K. Scott Wong and Sucheng Chan (Philadelphia: Temple University Press, 1998), 3–40; Alan M. Kraut, *Silent Travelers: Germs, Genes, and the "Immigrant Menace"* (New York: Basic Books, 1994).

16. David A. Hounshell, "Edison and the Pure Science Ideal in America," *Science* 207 (1980): 612. For further consideration of Rowland's "Plea," see "Comment and Criticism," *Science* 3 (29 February 1884): 241; Ronald Kline, "Construing 'Technology' As 'Applied Science': Public Rhetoric of Scientists and Engineers in the United States, 1880–1945," *Isis* 86 (1995): 200; and Daniel J. Kevles, *The Physicists: The History of a Scientific Community* (Cambridge, Mass.: Harvard University Press, 1995).

17. Rowland to his mother [Harriet Rowland], New Haven, 1868, Henry Augustus Rowland Papers, ms. 6, box 13, Special Collections, Johns Hopkins University [hereafter Rowland Papers].

18. *Henry Augustus Rowland* [biographical sketch], Rowland Papers, box 15. See also John David Miller, "H. A. Rowland and his Electromagnetic Researches" (Ph.D. diss., Oregon State University, 1970), 6; and Hall, *Life and Confessions*, 237.

19. William A. Anthony, "Address," *Proceedings of the American Association for the Advancement of Science* 36 (1887): 71, 77. For more on Anthony's response to Rowland's address, see Kline, "Construing 'Technology,'" 200.

20. Trowbridge to Gilman, 3 October 1883, Daniel Coit Gilman Papers, ms. 1, series 1, box 1.47, Special Collections, Johns Hopkins University [hereafter Gilman Papers]. Also see Wadsworth to Rowland, 1 September 1883, Rowland Papers, box 11. On other readers disgruntled by Rowland's "Plea," see Anthony, "Address," 71; and Kline, "Construing 'Technology,'" 200.

21. Gibbs to Gilman, 5 November 1883, Gilman Papers, ms. 1, box 1.16. "A Plea for Pure Science" was still being republished nearly twenty years after its initial delivery.

22. Henry Augustus Rowland, "A Plea for Pure Science," in *The Physical Papers of Henry Augustus Rowland* (Baltimore: Johns Hopkins University Press, 1902), 594, 595. Several advocates of pure science expressed the assumption that toil staved off degeneration.

G. Stanley Hall, for instance, often warned that societies would "relapse to barbarism" if not strengthened by rigorous education. See Hall, "New Departures in Education," *North American Review* 140 (February 1885): 151. Bederman explores Hall's ideas about racial recapitulation in *Manliness and Civilization*, 77–120.

23. Rowland, "Plea," 597–599, passim. References to the intellectual ascendance of Faraday, the "bookbinder's son," were widespread among American proponents of pure science. See Daniel Coit Gilman, "Inaugural Address," in *Addresses at the Inauguration of Daniel C. Gilman as President of the Johns Hopkins University* (Baltimore: Murphy, 1876), 61. On the cult of Faraday, see Geoffrey Cantor, "The Scientist As Hero: Public Images of Michael Faraday," in *Telling Lives in Science: Essays on Scientific Biography*, ed. Michael Shortland and Richard Yeo (Cambridge, U.K.: Cambridge University Press, 1996), 171–93.

24. See, for example, [W.H.W.], "Lawyers' Fees," *Central Law Journal* 11 (1880): 220.

25. Rowland, "Plea," 595, 608. Just a few years after penning these glowing paeans to lonely bachelorhood, the forty-one-year-old Rowland suggested that marriage would augment rather than detract from his scientific potency. Four days before his wedding to Henrietta Harrison in 1890, he described his hope that the marriage would "not only make my life more happy but also add to my working power in my science" (Rowland to Gilman, 1 June 1890, Gilman Papers, box 1.40).

26. Rowland, "Plea," 596, 605.

27. Daniel J. Kevles, "The Physics, Mathematics, and Chemistry Communities: A Comparative Analysis," in *Organization of Knowledge*, 141.

28. Rowland, "Plea," 606.

29. Jonathan Edwards, "Religious Affections," in *Representative Selections*, ed. Clarence H. Faust and Thomas H. Johnson (New York: Hill and Wang, 1962), 206.

30. Rowland, " Plea," 602, 598.

31. Ibid., 597; also see 604, 607. See also G. Stanley Hall, "Research the Vital Spirit of Teaching," *Forum* 17 (1894): 560. Similarly, the less accessible the knowledge was, the "higher" it became in the estimation of scientists: for example, "the most effectual way to lift the masses to a higher plane—materially, intellectually, morally—is to do everything favoring the climbing up of an ever increasing minority to higher and higher intellectual and moral altitudes" (Hermann E. von Holst, "Need of Universities in the United States," *Educational Review* 5 [1893]: 114).

32. Rowland, "Plea," 593.

33. Ibid., 596, 607.

34. "Presentation of the Rumford Medals to Professor Rowland," *Science* 3 (29 February 1884): 258.

35. Noyes, *Mastery of Submission*; Siegel, *Male Masochism*; Hall, "Confessions of a Psychologist," 120.

36. Rowland also recycled many of the speech's themes and images. See, for example, his "Electrical and Magnetic Discoveries of Faraday," in *Physical Papers*, 650–52; and his "The Highest Aim of the Physicist," *American Journal of Science and Arts* 8 (1899): 401–11.

37. John M. Coulter, *Mission of Science in Education* (Ann Arbor: University of Michigan, 1900), 7.

38. Hall, "Confessions of a Psychologist," 120.

39. Henry S. Carhart, "The Educational and Industrial Value of Science," *Science* 1 (12 April 1895): 397.

40. Henry S. Carhart, "The Humanistic Element in Science," *Science* 4 (31 July 1896): 125.

41. Hall, "Research the Vital Spirit," 559, 566. See also Richard von Krafft-Ebing, *Psy-*

chopathia Sexualis (New York: Putnam, 1965), 86ff. While Hall tended to describe the scientist as a devoted servant, occasionally he spoke instead of making nature an obedient slave to man—rhetoric particularly fraught in the years just after Reconstruction. See Hugh Hawkins, *Pioneer: A History of the Johns Hopkins University, 1874–1889* (Ithaca, N.Y.: Cornell University Press, 1960), 306.

42. Charles Sanders Peirce, "Pearson's *Grammar of Science*," in *The Essential Peirce: Selected Philosophical Writings*, ed. Nathan Houser et al. (Bloomington: Indiana University Press, 1998), 2:58.

43. Lorraine Daston, "Fear and Loathing of the Imagination in Science," *Daedalus* 127 (winter 1997): 90.

44. Quoted in ibid., 90.

45. Karl Marx, "Economic and Philosophical Manuscripts," in *Early Writings*, trans. Rodney Livingstone and Gregor Benton (London: Penguin, 1992 [1844]), 361.

46. Carhart, "Humanistic Element," 127. Changes in the concept of truth should not be overstated. Recent historical studies stress the continuity between late nineteenth-century research universities and the moral and spiritual aims of earlier denominational colleges. (See, for example, Robert E. Kohler, "The Ph.D. Machine: Building on the Collegiate Basis," *Isis* 81 [1990]: 638–62; Reuben, *Making of the Modern University*; and Guralnick, *Science and the Ante-Bellum American College*.) At the same time, Daston's work highlights a growing preoccupation with truth's mutability.

47. Thomas Edison, letter to *Baltimore Sun*, 16 April 1901, quoted in Hounshell, "Edison," 617, note 67. Also see "The Laboratory in Modern Science," *Science* 3 (15 February 1884): 173.

48. Hall, *Life and Confessions*, 236; Francesco Cordasco, *Daniel Coit Gilman and the Protean Ph.D: The Shaping of American Graduate Education* (Leiden: Brill, 1960), 75–76.

49. Veysey, *Emergence of the American University*, 152, 153. Osgood, a historian, was thoroughly aligned with the ethic of scientific self-sacrifice. As Peter Novick explains, no group of late nineteenth-century scholars "were more prone to scientific imagery, and the assumption of the mantle of science, than the historians" (33). For a fuller discussion of the relation between the historical profession and late nineteenth-century science, see Novick, *That Noble Dream: The "Objectivity Question" and the American Historical Profession* (Cambridge, U.K.: Cambridge University Press, 1988), 31ff.

50. Hall, *Life and Confessions*, 206, 238.

51. William Preston Johnston, *The Work of the University in America* (Columbia, S.C.: Presbyterian Publishing House, 1884), 14.

52. Quoted in Hawkins, *Pioneer*, 312.

53. Von Holst, "Need of Universities," 117.

54. Gilman to Miller, 5 January 1876, quoted in Hawkins, *Pioneer*, 91. On the freedom of universities, also see Hall, "Research the Vital Spirit," 570, 565; von Holst, "Need of Universities," 117; Johnston, *Work of the University*, 14; G. S. Morris, "University Education," *University of Michigan Philosophical Papers*, 1st ser., no. 1 (Ann Arbor, Mich.: Andrews and Witherby, 1886), 7, 9; and James Morgan Hart, *German Universities: A Narrative of Personal Experience* (New York: Putnam, 1874), 340.

55. Von Holst, "Need of Universities," 106.

56. The American interpretation of the German system bore no strict correlation to actual conditions, as Veysey describes (*Emergence of the American University*, 126). In this sense American scientists' revisionist memories might be compared to American physicians' changing recollections of their visits to France, as discussed by John Harley Warner in "Remembering Paris: Memory and the American Disciples of French

Medicine in the Nineteenth Century," *Bulletin of the History of Medicine* 65 (fall 1991): 301–25.

57. Friedrich Paulsen, *The German Universities and University Study*, trans. Frank Thilly and William W. Elwang (New York: Scribner, 1906), 266.

58. Von Holst, "Need of Universities," 106. On the relations between historians and science, see my note 49.

59. Owen Hannaway, "The German Model of Chemical Education in America: Ira Remsen at Johns Hopkins (1876–1913)," *Ambix* 23 (1976): 152.

60. Wadsworth to Rowland, 1 September 1883, Rowland Papers, box 11.

61. Hugh Hawkins describes the profound disunity of American sciences at this time in "University Identity: The Teaching and Research Functions," in *Organization of Knowledge*, 293–94. One indication of the anxiety experienced in this period of specialization was the proliferation of efforts to chart the divisions and branches of knowledge. See Lorraine Daston, "Academies and the Unity of Knowledge: The Disciplining of the Disciplines," *differences* 10, no. 2 (1998): 67–86.

62. Sally Gregory Kohlstedt, "Maria Mitchell and the Advancement of Women in Science," in *Uneasy Careers and Intimate Lives: Women in Science, 1789–1979*, ed. Pnina G. Abir-Am and Dorinda Outram (New Brunswick, N.J.: Rutgers University Press, 1987), 130–46.

63. Rossiter, *Women Scientists*, 45–46; Hawkins, *Pioneer*, 267–68; Veysey, *Emergence of the American University*, 133.

64. Rebecca Sharpe, "Random Remarks of a Lady Scientist," *Popular Science Monthly* 58 (1901): 548.

65. Helen Harelin Walworth, "Field Work by Amateurs," *Science* 1 (16 October 1880): 199.

66. Cited in Rossiter, *Women Scientists*, 14–15.

67. Quoted in John Hope Franklin and Alfred A. Moss, Jr., *From Slavery to Freedom: A History of African Americans*, 8th ed. (New York: Knopf, 2000), 301.

68. James M. Jay, *Negroes in Science: Natural Science Doctorates, 1876–1969* (Detroit: Balamp, 1971); James D. Anderson, "The Hampton Model of Normal School Industrial Education, 1868–1900," in *New Perspectives on Black Educational History*, ed. Vincent P. Franklin and James D. Anderson (Boston: Hall, 1978), 61–96; H. Kenneth Bechtel, "Introduction," in *Blacks, Science, and American Education*, ed. Willie Pearson, Jr., and H. Kenneth Bechtel (New Brunswick, N.J.: Rutgers University Press, 1989).

69. William Albert Noyes and James Flack Norris, *Biographical Memoir of Ira Remsen, 1846–1927* (Washington, D.C.: National Academy of Sciences, 1931); Frederick H. Getman, *The Life of Ira Remsen* (Easton, Pa: Journal of Chemical Education, 1940); Getman, "Ira Remsen: Erstwhile Dean of Baltimore Chemists," in Ira Remsen Papers, box 22, Special Collections, Johns Hopkins University [hereafter Remsen Papers]; Hannaway, "German Model," 152–53; and Hawkins, *Pioneer*, 60.

70. Veysey, *Emergence of the American University*, 171; Amy E. Tanner, "History of Clark University Through the Interpretation of the Will of the Founder" (1908), unpublished manuscript, Clark University Archives, Worcester, Mass.; Robert J. Kump, *G. Stanley Hall's Efforts to Implement the Humboldtian University Ideal at Clark University in Worcester, Massachusetts, USA* (Bern, Switzerland: Haupt, 1996).

71. Ira Remsen, "Is Science Bankrupt?" (1897), Remsen Papers, box 11. When asked by one visitor what he would do with his students, Rowland, without looking up from his work, allegedly replied, "Do with them? Do with them?—*I shall neglect them*" (Hawkins, *Pioneer*, 218). Also see Hall, *Life and Confessions*, 237. On Remsen's abilities as a teacher, see Hawkins, *Pioneer*, 60, 218; Getman, *Life of Ira Remsen*; Veazey to Reid, 10 June 1927, Remsen Papers, box 23; and Hannaway, "German Model," 153.

72. Ira Remsen, "The Significance of Chemical Laboratories" (1910), Remsen Papers, box 11.

73. Hart to Reid, 21 May 1927, Remsen Papers, box 23; Hannaway, "German Model," 154. See also Remsen's address "This is the Age of Science," in which he extols the poverty of Swedish apothecary Carl Scheele (Remsen Papers, box 11).

74. Dohme to Reid, 7 June 1927, Remsen Papers, box 23.

75. There was some dispute over whether Remsen or his student Constantine Fahlberg deserved credit for the discovery of saccharin. Fahlberg alone reaped the material reward for the work. See Hart to Reid, 18 May 1927 and 29 May 1927, Remsen Papers, box 23.

76. Reid to Derby, 30 May 1927, Remsen Papers, box 23; Hannaway, "German Model," 157; Hawkins, Pioneer, 140.

77. Ira Remsen, "Thoroughness" (1872), Remsen Papers, box 12; Shimomura to Reid, 24 July 1927, Remsen Papers, box 23; Hannaway, "German Model," 153.

78. "Reminiscences of Professor Remsen by C. E. Waters," Remsen Papers, box 23; Remsen, quoted in William A. Noyes, "Ira Remsen," Science 66 (16 September 1927): 246.

79. See, for instance, the advertisement for Andrew Jackson Davis's Sacred Gospels of Arabula tucked inside Remsen's small leather notebook, Remsen Papers, box 9.

80. Jerome Alexander, "Ira Remsen, '65" [obituary], City College Alumnus 25 (January 1929): 28, copy in Remsen Papers, box 23.

81. Ira Remsen, "All Members One of Another," Alumni 1 (November 1912), unpaginated, copy in Remsen Papers, box 1; Noyes, "Ira Remsen," 245. On Remsen's aversion to isolation, see his obituary in the Baltimore Sun (1927), copy in Remsen Papers, box 22.

82. Noyes, "Ira Remsen," 244. Also see Sophie Forgan, "The Architecture of Science and the Idea of a University," Studies in the History and Philosophy of Science 20, no. 4 (1989): 405–34.

83. Although one black student, Kelly Miller, studied briefly at Hopkins in the late 1880s, other black students were unofficially excluded from laboratories and seminars (Hall, Life and Confessions, 246). Hopkins student Walter H. Page noted that, despite their formal prohibition, a few white women attended lectures, including one young Quaker who "seems to keep apace with all that goes on." See Page to his cousin Sarah Jasper, 30 September 1876, reprinted in Burton J. Hendrick, The Training of an American: The Earlier Life and Letters of Walter H. Page, 1855–1913 (Boston: Houghton Mifflin, 1928), 75.

84. "Reminiscences." On smoking, see Howard to Reid, 16 December 1927, Remsen Papers, box 23; on shirt sleeves, see Hannaway, "German Model," 153–54.

85. Hart to Reid, 29 May 1927, Remsen Papers, box 23; Howard to Reid, 16 December 1927; Veysey, Emergence of the American University; Hannaway, "German Model"; Getman, Life of Ira Remsen, 68.

86. Remsen, "Is Science Bankrupt?"; Moale to Reid, 19 July 1927, Remsen Papers, box 23; Hannaway, "German Model," 154.

87. Holmes to Reid, 4 January 1928, Remsen Papers, box 23.

88. Lyons to Reid, 17 June 1927, Remsen Papers, box 23.

89. Friedrich Nietzsche, Daybreak: Thoughts on the Prejudices of Morality, trans. R. J. Hollingdale (Cambridge, U.K.: Cambridge University Press, 1982 [1881]), 16.

90. Hendrick, Training of an American, 73; Hawkins, Pioneer, 283.

91. Of the eighty-eight outstanding chemists included in the Dictionary of American Men of Science listed through 1910, twenty-four had been trained by Remsen. Johns Hopkins graduated twice as many of these chemists as any other American university (Getman, Life of Ira Remsen, 155). Also see Elliott, History of Science, 108; Hofstadter and Hardy, Development and Scope, 64; and Veysey, Emergence of the American University, chap. 3.

92. G. Brown Goode, "America's Relation to the Advance of Science," *Science* 1 (4 January 1895): 5.

93. Ellen Richards, "The Elevation of Applied Science to an Equal Rank with the So-Called Learned Professions," in *Technology and Industrial Efficiency* (New York: McGraw-Hill, 1911), 125. For a similar critique, see W. H. Walker, "Chemical Research and Industrial Progress: What Commerce Owes to Chemistry," *Scientific American Supplement* 72 (1 July 1911): 14–16.

CHAPTER 4   EXPLORERS

1. Frederick A. Cook, *My Attainment of the Pole: Being the Record of the Expedition that First Reached the Boreal Center, 1907–1909, With the Final Summary of the Polar Controversy* (New York: Kennerley, 1912), 43.

2. Theon Wright, *The Big Nail: The Story of the Cook-Peary Feud* (New York: Day, 1970), 22.

3. Fridtjof Nansen, *In Northern Mists: Arctic Exploration in Early Times* (New York: Stokes, 1911), 1:4.

4. See Urban Wråkberg, *Vetenskapens vikingatåg: Perspektiv på svensk polarforskning, 1860–1930* (Uppsala, Sweden: Institutionen för Idé-och Lärdomshistoria, 1995).

5. W. Gillies Ross, "Nineteenth-Century Exploration of the Arctic," in *North American Exploration*, ed. John Logan Allen (Lincoln: University of Nebraska Press, 1997), 3:298

6. Adolphus Washington Greely, *Three Years of Arctic Service: An Account of the Lady Franklin Bay Expedition of 1881–84* (New York: Scribner, 1886), 1:37; Doug Wilkinson, *Arctic Fever: The Search for the Northwest Passage* (Toronto: Clarke, Irwin, 1971), viii.

7. Cook, *Attainment*, 28, 42. For an example of the framing of the Arctic explorer's "great indefinable yearning for something unattainable" in explicitly sexual and romantic terms, see Donald B. MacMillan, *How Peary Reached the Pole* (Boston: Houghton Mifflin, 1934), 107. For fuller discussion of this issue, see Louis Montrose, "The Work of Gender in the Discourse of Discovery," *Representations* 33 (winter 1991): 1–41.

8. On the broader context of these representations, see Eric G. Wilson, *The Spiritual History of Ice: Romanticism, Science, and the Imagination* (New York: Palgrave Macmillan, 2003).

9. A. A. Milne, "In Which Christopher Robin Leads an Expotition to the North Pole," *Winnie-the-Pooh* (New York: Dell, 1978 [1926]), 110–29.

10. "The North Pole," *Century* 79 (November 1909): 152.

11. Matthew A. Henson, *A Negro Explorer at the North Pole* (New York: Arno/New York Times, 1969), 134–35. Arctic explorer Herbert M. Frisby, among others, has suggested that Henson was actually first to reach the pole, which would trouble Henson's published account of the end of the journey. See Frisby's introduction to Henson's *A Negro Explorer*, iv.

12. Cook, *Attainment*, 894.

13. S. Allen Counter, *North Pole Legacy: Black, White and Eskimo* (Amherst: University of Massachusetts Press, 1991), 209. Also see Annie Dillard, "An Expedition to the Pole," in *Surviving Crisis*, ed. Lee Gutkind (New York: Tarcher/Putnam, 1997); William H. Goetzmann, "A 'Capacity for Wonder': The Meanings of Exploration," in *North American Exploration*, 3:532–34.

14. Elsa Barker, "The Frozen Grail," in *The Little Book of Modern Verse*, ed. Jessie B. Rittenhouse (Boston: Houghton Mifflin, [1913]), 120. Barker, Peary's personal friend, apparently wrote the first account of his journey to the pole, which appeared in the magazine *Hampton's* under Peary's authorship. See the testimony of Lilian Eleanor Kiel

in *Congressional Record* 6284 (4 March 1915), copy in the Rear Admiral Robert E. Peary Family Collection, record group 401, box 2, National Archives and Records Administration, College Park, Maryland [hereafter Peary Collection].

15. Cook announced his attainment on 1 September 1909 but claimed to have reached the pole on 21 April 1908. See Wright, *Big Nail*, 2.

16. Dennis Rawlins, *Peary at the North Pole: Fact or Fiction?* (Washington, D.C.: Luce, 1973), 10.

17. See the undated letter to Cook [probably written 1909] signed by an unnamed "Cook-American" in the Frederick Albert Cook Papers, reel 2, Manuscript Division, Library of Congress, Washington D.C. [hereafter Cook Papers].

18. Karl Decker, "Dr. Frederick A. Cook—Faker," *Metropolitan* 31 (January 1910): 416–35. On Cook, see John Edward Weems, *Race for the Pole* (New York: Holt, 1960), 182.

19. Of 76,052 opinions gathered by one Pittsburgh paper in September 1909, 96 percent favored Cook. Of that number, 73,238 said Cook discovered the pole in 1908; 2,814 said Peary discovered the pole in 1909; 15,229 said Peary was second to the pole in 1909; 2,814 said Cook did not reach the pole in 1908; and 58,009 said Peary did not reach the pole in 1909. See Rawlins, *Peary at the North Pole*, 166.

20. Readers interested in a particularly thorough account might consult Robert Bryce's magisterial 1,133-page *Cook & Peary: The Polar Controversy, Resolved* (Mechanicsburg, Pa.: Stackpole, 1997).

21. Robert E. Peary, *Nearest the Pole* (London: Hutchinson, 1907), ix.

22. Quoted in Rawlins, *Peary at the North Pole*, 10.

23. Jeannette Mirsky, *To the North! The Story of Arctic Exploration from Earliest Times to the Present* (New York: Viking, 1934), 14. For other examples, see Richard Perry, *The Jeannette: And a Complete and Authentic Narrative Encyclopedia to the North Polar Region* (San Francisco: Roman, 1883), 19–28; Sir John Leslie, *Narrative of Discovery and Adventure in the Polar Seas and Regions* (Edinburgh: Oliver and Boyd, 1845), chap. 3; and Wright, *Big Nail*, 332.

24. Ross ties the sudden appearance of this ice to the eruption of Mount Tamboro in the Dutch East Indies in 1815. See his "Nineteenth-Century Exploration of the Arctic," 255.

25. Lisa Bloom rightly argues that polar expeditions, "far from being innocent of the tensions of empire," are "icons of the whole enterprise of colonialism." See her expert treatment of masculinity and imperialism in *Gender on Ice: American Ideologies of Polar Expeditions* (Minneapolis: University of Minnesota Press, 1993), 3.

26. The year 1818 also marked the publication of the first edition of Aaron Arrowsmith's "Map of Countries Round the North Pole," which provided unprecedented detail for explorers heading north of fifty degrees. See Ross, "Nineteenth-Century Exploration of the Arctic," 249ff.

27. With the advent of the Arctic craze, American attention shifted away from Antarctica, for reasons that could benefit from further investigation. For a discussion of American interest in Antarctica, see William E. Lenz, *The Poetics of the Antarctic: A Study in Nineteenth-Century Cultural Perceptions* (New York: Garland, 1995).

28. Mirsky, *To the North!*, 153.

29. "Sir John Franklin and the Arctic Regions," *North American Review* 71 (July 1850): 178.

30. Samuel M. Smucker, *Arctic Explorations and Discoveries during the Nineteenth Century* (New York: Lovell, 1886), 198.

31. Clayton to Lady Jane Franklin, 25 April 1849, quoted in ibid., 328.

32. For the search expeditions' influence on British exploration, see Ian R. Stone, "The Franklin Expedition in Parliament," *Polar Record* 32 (1996): 209–16.

33. Smucker, *Arctic Explorations*, v.

34. Ibid.

35. William Bradford, *The Arctic Regions: Illustrated with Photographs Taken on an Art Expedition to Greenland* (London: Sampson Low, Marston, Low, and Searle, 1873); Lenz, *Poetics of the Antarctic*, 94–95. For a subtle analysis of the relation among readers, writers, and explorers at this time, see Michael Robinson, *The Coldest Crucible: Arctic Exploration and American Culture, 1850–1910* (Chicago: University of Chicago Press, forthcoming).

36. W. Leaman, "The Scope and Spirit of Scientific Research," *Mercersburg Review* 20 (October 1873): 531, 529, 530; Hollinger, "Inquiry and Uplift." For a discussion of Victorian exploration as a masochistic undertaking, see Elaine Freedgood, *Victorian Writing about Risk: Imagining a Safe England in a Dangerous World* (Cambridge, U.K.: Cambridge University Press, 2000), especially chap. 4.

37. Chauncy C. Loomis, *Weird and Tragic Shores: The Story of Charles Francis Hall, Explorer* (New York: Knopf, 1971); William R. Hunt, *To Stand at the Pole: The Dr. Cook–Admiral Peary North Pole Controversy* (New York: Stein and Day, 1981), 72–73; C. H. Davis, ed., *Narrative of the North Polar Expedition: U.S. Ship Polaris, Captain Charles Francis Hall Commanding* (Washington, D.C.: U.S. Government Printing Office, 1876), 36.

38. Davis, *Narrative*, 17, 18, 20.

39. Ibid., 179, 43.

40. Ibid., 37, 27.

41. Ibid., 210.

42. Joseph Henry, 9 June 1871, in ibid., 638.

43. Ibid.

44. Rowland, "Plea," 608, emphasis added.

45. Karl Weyprecht, "Scientific Work of the Second Austro-Hungarian Polar Expedition, 1872–4," *Royal Geographical Society Journal* 45 (1875): 33.

46. Mirsky, *To the North!*, 14–15.

47. Weyprecht, quoted in G. E. Fogg, *A History of Antarctic Science* (New York: Cambridge University Press, 1992), 104; Baker, "First International Polar Year," 284, note 1.

48. Weyprecht, "Scientific Work," 33.

49. Fogg, *History of Antarctic Science*, 103ff; Baker, "First International Polar Year"; William Barr, "The Expeditions of the First International Polar Year, 1882–83," *The Arctic Institute of North America*, Technical paper 29 (Calgary, Alberta: Arctic Institute of North America, 1985), 6–45. On Weyprecht, see Greely, *Three Years of Arctic Service*, 19–24; and Barr, "Expeditions," 2–3.

50. The ship's crew explored the meanings of manhood even while they sat trapped in the ice: for example, during their long encumbrance in the frozen sea, the men on board entertained themselves with drag shows. Emma De Long, ed., *The Voyage of the Jeannette. The Ship and Ice Journals of George W. De Long, Lieutenant-Commander U.S.N., and Commander of the Polar Expedition of 1879–1881* (Boston: Houghton Mifflin, 1884), 2:495; Raymond Lee Newcomb, ed., *Our Lost Explorers: The Narrative of the Jeannette Arctic Expedition, as Related by the Survivors, and in the Records and Last Journals of Lieutenant De Long* (Hartford, Conn.: American Publishing Company, 1882), 285; and David Serlin, "Crippling Masculinity: Queerness and Disability in U.S. Military Culture, 1800–1945," *GLQ* 9, nos. 1 and 2 (2003): 149–79.

51. De Long, *Voyage of the Jeannette*, 869.

52. I employ the term *Eskimo* here in accordance with my early twentieth-century sources. Similarly, I am faithful to Henson's use of the term *Negro*. It seems a mistake to swap these terms for contemporary racial terminology without specific strategic reasons for

doing so. To suggest that a term such as *Caucasian* or *Negro* is synonymous with *white* or *African American* is to imply that race has some timeless, pre-social, physical character—a claim that the study of history unravels.

53. See the appendices to Greely, *Three Years of Arctic Service.*

54. Quoted in Pierre Berton, *The Arctic Grail: The Quest for the Northwest Passage and the North Pole, 1818–1909* (New York: Viking, 1988), 485.

55. Ibid.

56. Ibid.

57. Quoted in Peary, *Nearest the Pole*, vii–viii. See also Bederman, *Manliness and Civilization*, 193.

58. Heber D. Curtis describes some of these difficulties in "Navigation Near the Pole," *United States Naval Institute Proceedings* 65 (January 1939): 14.

59. MacMillan, *How Peary Reached the Pole*, 134.

60. Donald B. MacMillan, "Peary as a Leader," unpublished manuscript, Donald Baxter MacMillan Collection, box 16, George J. Mitchell Department of Special Collections and Archives, Bowdoin College Library, Brunswick, Maine [hereafter MacMillan Collection].

61. W. T. Stead, "Character Sketch and Interview, Dr. F. A. Cook," *Review of Reviews* 40 (October 1909): 331.

62. MacMillan, *How Peary Reached the Pole*, 96.

63. Bradley Robinson, *Dark Companion* (New York: McBride, 1947), 135.

64. The privations endured by explorers also included actions that readers now might dubiously regard as "punishing." One historian reports, for instance, that Peary's American descendants claim that the explorer had "been forced by local custom to engage in sexual relations with the indigenous women, including Ahlikahsingwah [the mother of his child]. That was a condition of Peary's association with the Eskimo villagers. . . . He had to have sex with the women before he could gain their confidence." See Counter, *North Pole Legacy*, 46.

65. Donald B. MacMillan, "The Real Value of Arctic Work," undated speech, MacMillan Collection, box 16.

66. Allusions to hardship do occasionally appear in the logs themselves. One afternoon's observations at Cape Morris Jessup, for instance, led MacMillan to note: "Depth of water the same as yesterday 91 fathoms. The ice and snow here are both in the very worst conditions. At times I have crawled on hands and knees. It was impossible to walk without falling. It is d—— discouraging work." See diary 7 (19 April–31 May 1909), Thursday, 13 May 1909, 2 A.M., MacMillan Collection.

67. Explorers in the north, like earlier Victorian diarists, kept journals with the full expectation that they would be seen by others; a journal was private only "insofar as it purported like a diary to reveal his inner thoughts." See Michael T. Bravo, "The Accuracy of Ethnoscience: A Study of Inuit Cartography and Cross-Cultural Commensurability," *Manchester Papers in Social Anthropology* 2 (1996): 8. On the general historiographical problem of experience, see Scott, "Experience."

68. See, for example, Peary Collection, record group 401, box 22.

69. MacMillan, *How Peary Reached the Pole*, 224.

70. Peary Collection, record group 401, box 7A.

71. Rawlins, *Peary at the North Pole*, 155.

72. "Legal Proof of the Discovery of the Pole," *Bench and Bar* 18 (July 1909): 87. My thanks to Tal Golan for bringing this article to my attention.

73. The record that Peary attempted to leave at the pole was discussed at length by Henry T. Helgesen, congressman from North Dakota, in the House of Representatives on 21 July

1916 and was published in the *Congressional Record* 13959 (3 August 1916, copy in Peary Collection, record group 401, box 2). Cook's description of the record he left at the pole appears in Stead, "Character Sketch," 338.

74. See [Nichols?] to Hubbard, 7 March 1910, Peary Collection, record group 401, box 71.

75. Daston and Galison, "Image of Objectivity," 120.

76. Although early twentieth-century commentators did not find the photographic gaze convincing, late twentieth-century critics returned to the polar photographs as adequate documentary evidence for proving or disproving the disputants' claims. See Bloom, *Gender on Ice*, 55–56.

77. See the folder marked "Peary Rebuttal Material," Peary Collection, record group 401, box 3.

78. MacMillan, *How Peary Reached the Pole*, 283.

79. See, for example, *Statement of Captain Robert E. Peary, U.S.N. to Subcommittee No. 8 of the Committee on Naval Affairs*, U.S. House of Representatives, 7 January 1911, p. 11, copy in Peary Collection, record group 401, box 1.

80. Karen Halttunen, *Confidence Men and Painted Women: A Study of Middle-Class Culture in America* (New Haven, Conn.: Yale University Press, 1982).

81. See, for example, Morris Fishbein, *The New Medical Follies; An Encyclopedia of Cultism and Quackery in these United States* (New York: Boni and Liveright, 1927); and Frank Spencer, ed., *Piltdown Papers: 1908–1955* (London: Oxford University Press, 1990).

82. In this sense, the explorer's suffering provided a "technology of trust" akin to those described by historians of experiment Shapin and Schaffer in *Leviathan and the Air-Pump*, 60.

83. MacMillan, *How Peary Reached the Pole*, 290.

84. Even before his polar success in 1909, Peary appeared in the popular press as the embodiment of a physically powerful, manly integrity. See, for example, "Peary Likely to be a Rich Man," *Boston Sunday Globe*, 23 February 1908, p. 12.

85. Cook, *Attainment*, 287.

86. Stead, "Character Sketch," 336.

87. [Evening Star?], 11 September 1909, copy in Peary Collection, record group 401, box 186.

88. Henson, *Negro Explorer*.

89. Bridgman to Henson, 18 October 1909: "I think you did first-class last night, only don't let them tangle you up about observations, and fool questions about what you thought before you left about what Cook was going to do. Mr. Brady [sic] can tell the audience to stick to questions (straight) which can be answered, and don't involve guesswork by you" (Peary Collection, record group 401, box 7).

90. "Legal Proof," 88.

91. Quoted by Helgesen; see note 76.

92. "Legal Proof," 88.

93. Ibid.

94. "Professor Penck on the Polar Controversy," *Popular Mechanics* (December 1909): 786–87.

95. On rare occasions, Cook drew attention to the suffering of the Eskimos (see *Attainment*, 270, 279); but in each case, their suffering was presented as evidence of their cultural and racial status, not of their contributions to knowledge.

96. MacMillan, *How Peary Reached the Pole*, 62.

97. See, for example, the description of Kood-look-to's "astonishing" curiosity in ibid., 249–52.

98. Peary, *Nearest the Pole*, 375. He asserts that civilized communities would see the Eskimos' objects of fascination as "rubbish" (376).

99. Ibid., 390.

100. See "Explorer," 16 June 1934, clipping in Peary Collection, record group 401, box 185. Eventually, Henson was awarded a $2,500 pension; see U.S. House of Representatives 9988, 72d Congress, 1st session, "A Bill Granting a Pension to Matthew A. Henson," 1 March 1932, copy in Peary Collection, record group 401, box 185.

101. Josephine Diebitsch Peary, *My Arctic Journal: A Year Among Ice-Fields and Eskimos* (New York: Contemporary, 1893). In a heartbreaking set of diary entries from 1893, Josephine Peary voices the loneliness, fatigue, and "fit of the blues" she suffered after the birth of her daughter, Marie. See JDPS18 [1893 diary, 10 September–12 November] in the Josephine Diebitsch Peary Collection, Maine Women Writers Collection, University of New England, Portland, Maine.

102. On white women's complex roles as writers, readers, and explorers in expedition narratives, see Dea Birkett, *Spinsters Abroad: Victorian Lady Explorers* (New York: Blackwell, 1989); Bloom, *Gender on Ice*, 41; and Robinson, *Coldest Crucible*. On Peary's attitudes toward women, see John Edward Weems, *Peary: The Explorer and the Man* (Boston: Houghton Mifflin, 1967), 71–72. On the necessity of a feminized audience to receive the stories of manly heroism in another context, see Mary Terrall, "Gendered Spaces, Gendered Audiences: Inside and Outside the Paris Academy of Sciences," *Configurations* 2 (1995): 223.

103. Compare Bederman, *Manliness and Civilization*, 77–120.

104. MacMillan, *How Peary Reached the Pole*, 96. On how "Peary's admirable organization" of dogs, men, and equipment was designed to minimize risk and suffering, see Cyrus C. Adams, "The North Pole at Last," *American Review of Reviews* 40 (1909): 422. For a comparision of Peary's and Taylor's systems see MacMillan, *How Peary Reached the Pole*, 289–90. On Peary's re-engineering of men, machines, and dogs into scientific instruments, see Bloom, *Gender on Ice*, 44–46.

105. While I follow Bloom in stressing the importance of imperial dominance in Arctic exploration, her account neglects several of the more interesting contradictions and internal tensions in these positions of dominance. When the *Maine* blew up in Havana Harbor in February 1898, for instance, Peary grew concerned that a war would hinder his efforts to reach the pole. As he wrote to his wife, "Well, Jo, last time they staged a silver panic to keep me from going [north]. Now they're going to have a war!" (quoted in Rawlins, *Peary at the North Pole*, 38). Here, as elsewhere, the uncommonly privileged and powerful liberal subject positions himself as victim. Henson's experiences reveal similar complexity; for a fuller discussion, see Counter, *North Pole Legacy*.

106. MacMillan, *How Peary Reached the Pole*, 183.

107. Ibid., 184.

108. Peary Collection, record group 401, box 186.

109. [Edina Alonzo Bletke?], 11 September 1909, Cook Papers, reel 2.

110. MacMillan, *How Peary Reached the Pole*, 61.

111. "A Moral or Two from the Polar Controversy," *Century* 79 (March 1910): 793.

112. "E.S.V.Z.," "Rime of the Modern Mariner," copy in Peary Collection, record group 401, box 7.

113. W. L. Gage, "Introduction," in Newcomb, *Our Lost Explorers*, iii.

114. Erich Maria Remarque, *All Quiet on the Western Front* (New York: Ballantine, 1989 [1928]), 273.

115. The reliability of exploratory knowledge in other contexts has been assured through techniques both similar to and different from those used here. Compare Dorinda Outram, "On Being Perseus: Travel and Truth in the Englightenment," in *Geography and*

the Enlightenment, ed. David N. Livingstone and Charles W. J. Withers (Chicago: University of Chicago Press, 1999), 281–94; D. Graham Burnett, *Masters of All They Surveyed: Exploration, Geography, and a British El Dorado* (Chicago: University of Chicago Press, 2000), 99–117; Mary Terrall, "Heroic Narratives of Quest and Discovery," *Configurations* 6 (1998): 223–42; and Bruce Hevly, "The Heroic Science of Glacier Motion," *Osiris* 11 (1996): 66–86.

CHAPTER 5   MARTYRS

1. See, for example, "Untoward Effects of X-rays," *Boston Medical and Surgical Journal* 152 (1905): 173; and S. Burt Wolbach, "Summary of the Effects of Repeated Roentgen-Ray Exposures Upon the Human Skin, Antecedent to the Formation of Carcinoma," *American Journal of Roentgenology and Radiology* 13 (1925): 139–43.

2. "Operated on 72 Times: Roentgenologist Has Lost Eight Fingers and an Eye for Science," *New York Times*, 12 March 1926, p. 22. See also "Undergoes 50th Operation: Dr. Baetjer of Johns Hopkins is a Victim of X-Ray Infections," *New York Times*, 2 May 1924, p. 26.

3. Percy Brown, *American Martyrs to Science through the Roentgen Rays* (Springfield, Ill.: Thomas, 1936), 98.

4. Sande Bishop, *Radiology in New England: The First Hundred Years* (Boston: Massachusetts Radiological Society/New England Roentgen Ray Society, 1995), 74.

5. John V. Pickstone, *Ways of Knowing: A New History of Science, Technology, and Medicine* (Manchester, U.K.: Manchester University Press, 2000), 48. Historians of radiology have joined early roentgenologists in stressing the enthusiasm of early researchers. According to one historian, "the answer" to why so many X-ray investigators cheerfully subjected themselves to manifestly life-threatening work "seems to be largely enthusiasm" (Stephen B. Dewing, *Modern Radiology in Historical Perspective* [Springfield, Ill.: Thomas, 1962], 83). "They were fully aware of what awaited them; fully aware," writes another historian, calling the experimenters "martyrs without a frontier." X-ray researchers "paid no heed" to matters such as cancer, for "hunting down shadows was far too fascinating a pursuit for them to be induced to slacken the pace" (Pino Donizetti, *Shadow and Substance: The Story of Medical Radiography*, trans. Anne Ellis [Oxford: Pergamon, 1967], 147–48).

6. Like most people engaged in a novel endeavor, early practitioners of roentgenology debated nomenclature. I here adopt the terms *roentgenology* and *roentgenologist* in favor of today's more familiar *radiology* in adherence to the preferences of the first relevant professional body in the United States, the American Roentgen Ray Society. *Radiology*, moreover, may connote any study of high-energy radiation, while *roentgenology* refers specifically to the field of X rays. I use the anglicized spelling *Roentgen* for the German scientist's name and the alternate *Röntgen* only when directly citing period sources.

7. For recent treatments of nationalism that focus on the significance of sacrifice, see Michael Rowlands, "Memory, Sacrifice, and the Nation," *New Formations* 30 (winter 1996): 8–17; and Adam J. Lerner, "The Nineteenth-Century Monument and the Embodiment of National Time," and Elshtain, "Sovereignty, Identity, Sacrifice," both in *Reimagining the Nation*.

8. The professionalization of X-ray work has been described at length by others. I here focus only on the centrality of martyrdom in this process. Other influential accounts of the rise of roentgenology include Arthur C. Christie, "Fifty Years of Progress in Ra-

diology," in *The American Roentgen Ray Society, 1900–1950: Commemorating the Golden Anniversary of the Society* (Springfield, Ill: Thomas, 1950), 25–29; Arthur U. Desjardins, "The Status of Radiology in America," *Journal of the American Medical Association* 92 (1929): 1035–39; Joel D. Howell, *Technology in the Hospital: Transforming Patient Care in the Early Twentieth Century* (Baltimore: Johns Hopkins University Press, 1995), especially 103–32; George Sarton, "The Discovery of the X-Rays," 26 *Isis* (1937): 340–69; Nancy Knight, "The New Light: X-Rays and Medical Futurism," in *Imagining Tomorrow: History, Technology, and the American Future*, ed. Joseph Corn (Cambridge, Mass.: MIT Press, 1986); Ronald L. Eisenberg, *Radiology: An Illustrated History* (St. Louis, Mo.: Mosby, 1992); E.R.N. Grigg, *The Trail of the Invisible Light* (Springfield, Ill.: Thomas, 1965); and Stanley Joel Reiser, *Medicine and the Reign of Technology* (Cambridge, U.K.: Cambridge University Press, 1978), 58–60. Canonical narratives of Roentgen's discovery have been scrutinized only rarely; for two critiques, see Lisa Cartwright, *Screening the Body: Tracing Medicine's Visual Culture* (Minneapolis: University of Minnesota Press, 1995); and Thomas S. Kuhn, *The Structure of Scientific Revolutions*, 2d ed. (Chicago: University of Chicago Press, 1970 [1962]), 57–59.

9. Robert G. Arns, "The High-Vacuum X-Ray Tube: Technological Change in Social Context," *Technology and Culture* 38 (October 1997): 852–90; Merrill C. Sosman, "Roentgenology at Harvard," *Harvard Medical Alumni Bulletin* 21 (April 1947): 65–76.

10. John Daniel, "The X-rays," *Science* 3 (10 April 1896): 562–63.

11. "Deleterious Effects of X Rays on the Human Body," *Electrical Review* 29 (12 August 1896): 78; Bettyann Holtzmann Kevles, *Naked to the Bone: Medical Imaging in the Twentieth Century* (New Brunswick, N.J.: Rutgers University Press, 1997), 47.

12. G. A. Frei, "Deleterious Effects of X Rays on the Human Body: Further Evidence that Repeated Exposure to the Rays Produces a Sunburn Effect," *Electrical Review* 29 (19 August 1896): 95.

13. Fred S. Jones, "Deleterious Effects of X Rays on the Human Body," *Electrical Review* 29 (9 September 1896): 127.

14. S.J.R., "Some Effects of the X-Rays on the Hands," *Nature* 54 (29 October 1896): 621.

15. H. D. Hawks, "The Physiological Effects of the Roentgen Rays," *Electrical Engineer* 22 (16 September 1896): 276.

16. W. C. Fuchs, "Effect of Röntgen Rays on the Skin," *Western Electrician* 19 (12 December 1896): 291; E. A. Codman, "The Cause of Burns from X-Rays," *Boston Medical and Surgical Journal* 135 (10 December 1896): 610; W. M. Stine, "Effect on the Skin of Exposure to the Roentgen Tubes," *Electrical Review* 29 (18 November 1896): 250; Leonard, quoted in Kevles, *Naked to the Bone*, 47; William Harvey King, *Electricity in Medicine and Surgery, Including the X-ray* (New York: Boericke and Runyon, 1901), 114; Williams, quoted in Ruth Brecher and Edward Brecher, *The Rays: A History of Radiology in the United States and Canada* (Baltimore: Williams and Wilkins, 1969), 86. Also see Arns, "High-Vacuum X-Ray Tube."

17. There were other famous debates in professional publications over the source of the burns, notably between William Herbert Rollins and Ernest Codman. On this debate, see Bishop, *Radiology in New England*, 51–53.

18. G. A. Frei, "X-Ray Physiological Effects," *Electrical Engineer* 22 (16 September 1896): 276; Frei, "X-Rays Harmless with the Static Machine," *Electrical Engineer* 22 (23 December 1896): 651; Frei, "X-Rays Harmless with the Static Machine," *Electrical World* 29 (2 January 1897): 27.

19. Elihu Thomson, "Some Recent Röntgen-Ray Work," *Electrical World* 28 (10 October 1896): 415.

20. David O. Woodbury, *Elihu Thomson: Beloved Scientist* (Boston: Museum of Science, 1960), 221.

21. Kevles, *Naked to the Bone*, 49; Daniel Paul Serwer, "The Rise of Radiation Protection: Science, Medicine, and Technology in Society, 1896–1935" (Ph.D. diss., Princeton University, 1977), 48; Bishop, *Radiology in New England*, 21.

22. Elihu Thomson, "Some Notes on Roentgen Rays," *Electrical Engineer* 22 (18 November 1896): 520–21; Thomson "Röntgen Rays Act Strongly on the Tissues," published in *Electrical World* 28 (28 November 1896): 666, *Electrical Review* 29 (25 November 1896): 260, and *Electrical Engineer* 22 (25 November 1896): 534.

23. Thomson, "Roentgen Rays Act Strongly on the Tissues" (*Electrical Engineer*), 534.

24. Stine, "Effect on the Skin," 250.

25. See, for example, Codman, "Cause of Burns."

26. R. B. Owens, "Effect of Röntgen Rays on the Tissues," *Electrical World* 28 (19 December 1896): 759.

27. Greene to Thomson, 20 December 1896, in Harold J. Abrahams and Marion B. Savin, eds., *Selections from the Scientific Correspondence of Elihu Thomson* (Cambridge, Mass.: MIT Press, 1971), 247.

28. Philip Atkinson, *Electricity for Everybody; Its Nature and Uses Explained*, 2d ed. (New York: Century, 1897), 239 (the edition had been expanded to include a chapter on the new rays); Samuel Howard Monell, *A System of Instruction in X-Ray Methods and Medical Uses of Light, Hot-Air, Vibration, and High Frequency Currents* (New York: Pelton, 1902), 26.

29. Cited in Hector Alfred Colwell and Sidney Russ, *X-ray and Radium Injuries: Prevention and Treatment* (London: Oxford University Press, 1934), 9. On the period of incubation, see Thomson, "Roentgen Rays Act Strongly on the Tissues" (*Electrical Engineer*), 534; on patients' lesions, see J. C. White, "Dermatitis Caused by X-Rays," *Boston Medical and Surgical Journal* 135 (3 December 1896): 583.

30. On litigation, see Serwer, "Rise of Radiation Protection," 39–40. For a broader discussion of relationships between X rays and the law at this time, see Tal Golan, *Laws of Men and Laws of Nature: The History of Scientific Expert Testimony in England and America* (Cambridge, Mass.: Harvard University Press, 2004), chap. 5.

31. Carl Beck, *Röntgen Ray Diagnosis and Therapy* (New York: Appleton, 1904), 368ff; Brown, *American Martyrs*, 33, 40–41; Brecher and Brecher, *Rays*, 163.

32. "Edison and X-Ray Injuries," *Journal of the American Medical Association* 41 (22 August 1903): 499; Kevles, *Naked to the Bone*, 47–48.

33. "Death of a Famous Woman Radiographer," *San Francisco Chronicle*, 5 August 1905, p. 10. See also Peter E. Palmquist, *Elizabeth Fleischmann: Pioneer X-Ray Photographer* (Berkeley, Calif.: Magnes Museum, 1990).

34. "The Woman Who Takes the Best Radiographs in the World," *San Francisco Chronicle*, 3 June 1900, p. 30. The role of roentgenology in the war is detailed in W. C. Borden, *The Use of the Röntgen Ray by the Medical Department of the United States Army in the War with Spain, 1898* (Washington, D.C.: U.S. Government Printing Office, 1900); Lawrence Reynolds, "The History of the Use of the Roentgen Ray in Warfare," in *Classic Descriptions in Diagnostic Roentgenology*, ed. André J. Bruwer (Springfield, Ill: Thomas, 1964), 2:1307–17.

35. Brown, *American Martyrs.*

36. Sir David Brewster, *The Martyrs of Science* (New York: Harper, 1841); Tissandier, *Les martyrs.*

37. Henry Adams, *The Education of Henry Adams* (New York: Vintage, 1990 [1907]), 355.

38. Greene to Thomson, 20 December 1896, in *Selections*, 247–48.

39. "The Edison Fluoroscope Exhibit," *Electrical Engineer* 21 (3 June 1896): 600, 601.

40. Quoted in Otto Glasser, *Wilhelm Conrad Röntgen and the Early History of Röntgen Rays* (Springfield, Ill.: Thomas, 1934), 293–94.

41. Kevles, *Naked to the Bone*, 48.

42. Emil H. Grubbé, *X-Ray Treatment: Its Origin, Birth and Early History* (Saint Paul, Minn.: Bruce), 89, 79.

43. Grubbé claimed to have been the first experimenter in the world to develop and discuss X-ray sequelae (ibid., 45), but historians of radiology typically refute his assertion. See, for example, Brecher and Brecher, *Rays*, 91ff; and Paul C. Hodges, *The Life and Times of Emil H. Grubbé* (Chicago: University of Chicago Press, 1964).

44. Grubbé, *X-Ray Treatment*, 79.

45. Ralph Waldo Emerson, "Compensation," in *Selected Writings of Emerson*, ed. Donald McQuade (New York: Random House, 1981), 162.

46. Donizetti, *Shadow and Substance*, 148.

47. Lewis Gregory Cole, "In Memoriam," *American Journal of Roentgenology* 4 (1917): 630.

48. P. M. Hickey, "The First Decade of American Roentgenology," *American Journal of Roentgenology* 20 (1928): 150.

49. Charles Allen Porter, "The Pathology and Surgical Treatment of Chronic X-Ray Dermatitis," *Transactions of the American Roentgen Ray Society* 9 (1908): 128, 169–70; Hickey, "First Decade," 155; Brown, *American Martyrs*, 123.

50. Charles Lester Leonard, "The Protection of the Roentgenologist," *New York Medical Journal* 86 (16 November 1907): 917–20. See also Leonard, "The X-Ray 'Burn': Its Productions and Prevention. Has the X Ray any Therapeutic Properties?" *New York Medical Journal* 68 (2 July 1898): 18–20.

51. Brown, *American Martyrs*, 123.

52. Ibid., 158–59.

53. Mihran K. Kassabian, "X-Ray as an Irritant," *American X-Ray Journal* 7 (1900): 784–86.

54. Brown, *American Martyrs*, 93.

55. Ibid., 97–98.

56. Kevles, *Naked to the Bone*, 48; Grubbé, *X-Ray Treatment*, 139–40.

57. Alan Trachtenberg, *The Incorporation of America: Culture and Society in the Gilded Age* (New York: Hill and Wang, 1982), 91.

58. John Macy, *Walter James Dodd: A Biographical Sketch* (Boston: Houghton Mifflin, 1918), 11.

59. [Percy Brown?], "Walter James Dodd, 1869–1916," unpublished, undated manuscript, in manuscripts and letters regarding Walter J. Dodd (B MS C42.5), p. 1, Francis A. Countway Library of Medicine, Boston.

60. Porter, "Pathology and Surgical Treatment," 141; Macy, *Dodd*, 21, 23.

61. Macy, *Dodd*, 26.

62. Ibid., 143, 145.

63. Discussion of the white male body as a machine can be found in Anson Rabinbach, *The Human Motor: Energy, Fatigue, and the Origins of Modernity* (Berkeley: University of California, 1990); Roxanne Panchasi, "Reconstructions: Prosthetics and the Rehabilitation of the Male Body in World War I France," *differences* 7 (1995): 110–40; Bill Brown, "Science Fiction, the World's Fair, and the Prosthetics of Empire, 1910–1915," in *Cultures of United States Imperialism*, 129–63; Lisa Herschbach, "Prosthetic Reconstructions: Making the Industry, Re-making the Body, Modeling the Nation," *History Workshop Journal* 44 (1997): 23–57; and Siegfried Giedion, *Mechanization Takes Command: A Contribution to Anonymous History* (New York: Oxford University Press, 1948).

64. Macy, *Dodd*, 58.

65. Ibid., 29. Porter reports administering opiates to Dodd on one occasion ("Pathology and Surgical Treatment," 144).

66. Porter, "Pathology and Surgical Treatment," 159; Macy, *Dodd*, 29.

67. Macy, *Dodd*, 26; Sosman, "Roentgenology at Harvard," 71. I have been unable to determine if Dodd would have been eligible for any form of disability compensation if he had left his post at the hospital.

68. Harris P. Mosher, quoted in Sosman, "Roentgenology at Harvard," 69.

69. Macy, *Dodd*, 27.

70. Brown, *American Martyrs*, 149–50.

71. Only one scholar appears to have used the term *masochism* when describing the actions of the so-called martyrs: see Cartwright, *Screening the Body*, 110, 128.

72. Sosman, "Roentgenology at Harvard," 68.

73. Cole, "In Memoriam," 630.

74. [Brown?], "Dodd," 6.

75. Macy, *Dodd*, 35.

76. Ibid., 36.

77. Cole, "In Memoriam," 630.

78. George E. Pfahler, "Fifty Years of Trials and Tribulations in Radiology," in *American Roentgen Ray Society*, 17.

79. Brown, *American Martyrs*, 16.

80. Ibid.

81. Brown's emphasis on the special character of work echoes usage found in contemporary artistic and literary circles (such as photographer Alfred Stieglitz's journal *Camera Work*) and contributes to the aura of unprofitable devotion permeating both science and aesthetics. On transformations in the concept of work at this time, see Raymond Williams, *Keywords* (New York: Oxford University Press, 1976), 334–37.

82. Serwer, "Rise of Radiation Protection," 63.

83. Kevles, *Naked to the Bone*, 47.

84. Ibid., 48.

85. Political theorist Wendy Brown discusses the place of injury in the constitution of political identities in late modernity in *States of Injury: Power and Freedom in Late Modernity* (Princeton, N.J.: Princeton University Press, 1995), 52–76.

86. Nicola Tesla, "Tesla on the Hurtful Actions of Lenard and Roentgen Tubes," *Electrical Review* 30 (5 May 1897): 211.

87. This monument was part of a broader post–World War I remembrance of radiology, which sought not only to carry the X ray's wartime prominence into the peacetime promotion of radiology but also to reunite (symbolically and practically) an international field divided by the war. Brown's 1936 *American Martyrs* also contributed to this effort, but in a more nationalistic vein. For a lucid general discussion of the significance of memorials, portraits, and other forms of commemoration in understandings of science, see Ludmilla Jordanova, "Presidential Address: Remembrance of Science Past," *British Journal for the History of Science* 33 (2000): 387–406.

88. Rowlands, "Memory, Sacrifice, and the Nation," 11.

89. Bishop, *Radiology in New England*, 39–42; Jacalyn Duffin and Charles R. R. Hayter, "Baring the Sole: The Rise and Fall of the Shoe-Fitting Fluoroscope," *Isis* 91 (2000): 260–82. On the numerous other injuries caused by X radiation at the time, see Rebecca Herzig, "Removing Roots: 'North American Hiroshima Maiden Syndrome' and the X-Ray," *Technology and Culture* 40 (October 1999): 723–45.

90. See, for example, S. H. Sharpsteen, "The History of an X-ray Burn," *Electrical Engineer* 24 (8 July 1897): 10–11; and the discussion of William Herbert Rollins's experiments with guinea pigs in Kevles, *Naked to the Bone*, 51.

91. Brown, *American Martyrs*, 4; "Death of a Famous Woman Radiographer"; "Woman Who Takes the Best Radiographs."

92. On the ways in which representations of Curie's sacrifice strengthened popular conceptions of appropriate womanhood, see Susan Lindee, "The Scientific Romance: Purity, Self-Sacrifice, and Passion in Popular Biographies of Marie Curie," paper presented at the annual meeting of the History of Science Society, Atlanta, 8 November 1996; Helena M. Pycior, "Marie Curie's 'Anti-Natural Path': Time Only for Science and Family," in *Uneasy Careers*, 213. Poet Adrienne Rich explores Curie's wounds in "Power," in *The Dream of a Common Language, Poems 1974–1977* (New York: Norton, 1978), 3.

93. Brown, *American Martyrs*, 49. More recently, Palmquist elaborates on the exceptional achievements of Fleischmann-Ascheim, a young Jewish woman who apparently never completed high school, by comparing her accomplishments (and heroic sacrifices) to that other famously exceptional woman in science, Madame Curie. Again, Fleischmann-Ascheim's exceptionality is preserved, maintaining the general masculinization of the field. See Palmquist, *Fleischmann*, 10.

94. Brown, *American Martyrs*, 264.

95. On some of these other dangers, see Sosman, "Roentgenology at Harvard," 66–67.

96. Dewing, *Modern Radiology*, 75; Kevles, *Naked to the Bone*, 110.

CHAPTER 6   BARBARIANS

1. Sinclair Lewis, *Arrowsmith* (New York: New American Library, 1980 [1925]), 430. All further citations from the novel appear parenthetically in the text.

2. Paul de Kruif, *The Sweeping Wind: A Memoir* (New York: Harcourt, Brace and World, 1962), 116. Unfortunately, de Kruif's recollection of this conversation cannot be compared to Lewis's because the novelist's papers do not recall it in such detail.

3. The film version of *Arrowsmith* (directed by John Ford, 1931) accentuates the link between Arrowsmith's scientific and sexual unproductivity. Leora miscarries while Martin is away pursuing fruitless scientific research, implying that he might have been able to save the pregnancy had he been present. In the novel, on the other hand, Martin is with Leora when the dead fetus is removed. On the film, see Susan E. Lederer, "Film Review: *Arrowsmith*," *Isis* 84 (December 1993): 771–72; and Lisa L. Lynch, "*Arrowsmith* Goes Native: Medicine and Empire in Fiction and Film," *Mosaic* 33 (December 2000): 193–208.

4. See Rosenberg's chapter "Martin Arrowsmith: The Scientist As Hero," in *No Other Gods*, 279, note 14. See also Steven Shapin, "The Philosopher and the Chicken: On the Dietetics of Disembodied Knowledge," in *Science Incarnate*, 42; and Lynch, "*Arrowsmith* Goes Native."

5. Mark Schorer proposes that the ending emerges from Lewis's "sentimental notions about the heroic life of untrammeled nature, of nature's noblemen," and wonders aloud "if Paul de Kruif approved of the ending." See Schorer's "Afterword," in *Arrowsmith*, 435–36.

6. On the relations among Gottlieb, Loeb, and Novy, see Robert L. Coard, "Sinclair Lewis, Max Gottlieb, and Sherlock Holmes," in *Sinclair Lewis's "Arrowsmith,"* ed. Harold Bloom (New York: Chelsea House, 1988); Rosenberg, "Martin Arrowsmith"; and Ilana Löwy,

"Immunology and Literature in the Early Twentieth Century: *Arrowsmith* and *The Doctor's Dilemma*," *Medical History* 32 (1988): 320–25.

7. Lewis to Harcourt, 21 September 1923, in Sinclair Lewis, *From Main Street to Stockholm: Letters of Sinclair Lewis, 1919–1930*, ed. Harrison Smith (New York: Harcourt Brace, 1952), 139.

8. Although Schorer notes that Lewis's heroes typically triumph "after sacrifice" ("Afterword," 437), I have trouble interpreting Lewis's other famous characters as triumphant. Carol Kennicott, in particular, succumbs unambiguously to the oppressive normalcy of Gopher Prairie. Arrowsmith, however, is distinguished by "his devotion to [the] *religion of science*," as Lewis summarized in his notebook while planning the novel. See Richard Lingeman, *Sinclair Lewis: Rebel from Main Street* (New York: Random House, 2002), 224.

9. Frederic I. Carpenter, "Sinclair Lewis and the Fortress of Reality," in *Sinclair Lewis's "Arrowsmith,"* 12. See also Rosenberg, "Martin Arrowsmith"; and Harold Bloom, "Introduction," and Mary G. Land, "Three Max Gottliebs: Lewis's, Dreiser's, and Walker Percy's View of the Mechanist-Vitalist Controversy," both in *Sinclair Lewis's "Arrowsmith."* Each author compares Arrowsmith's Vermont hideaway to Thoreau's Walden cabin. On the significance of nature as a purifying force in American literature more generally, see Miller, *Errand*, 204; Leo Marx, *The Machine in the Garden: Technology and the Pastoral Ideal in America* (New York: Oxford University Press, 1964); and Roderick Nash, *Wilderness and the American Mind* (New Haven, Conn.: Yale University Press, 1967).

10. Arthur Margon, "Changing Models of Heroism in Popular American Novels, 1880–1920," *American Studies* 17 (fall 1976): 71–86.

11. Nathaniel Hawthorne, "The Birthmark" and "Rappacini's Daughter," in *Hawthorne's Short Stories*, ed. Newton Arvin (New York: Vintage, 1946); Edward Bellamy, *Doctor Heidenhoff's Process* (New York: AMS Press, 1969 [1880]); and Jack London, "A Thousand Deaths," *Black Cat* 4 (May 1899): 33–42. For further discussion of scientists in English-language fiction, see Haynes, *From Faust to Strangelove*.

12. Rowland, "Plea," 597.

13. Lewis to Alf and Don, 10 April 1923, in Lewis, *From Main Street to Stockholm*, 130. Harcourt suggested that the title *Barbarian* might be too satirical, something Lewis sought to avoid after critics complained that he could only write satirically (Harcourt to Lewis, 23 April 1923, in ibid., 131). Lewis replied, "I *think* BARBARIAN, sans the article, is the title, but as you say, we must mull over it—" (Lewis to Harcourt, 3 May 1923, in ibid., 132).

14. On the difference between atavism and degeneracy, see Cynthia Eagle Russett, *Sexual Science: The Victorian Construction of Womanhood* (Cambridge, Mass.: Harvard University Press, 1989), 66–70.

15. The relation between madness and scientific genius is also developed in the figure of Gottlieb: for example, "It is possible that Max Gottlieb was a genius. Certainly he was as mad as any genius" (124).

16. Describing Lewis's impulsiveness and nervous exhaustion, Martin Light notes that real medical researchers could not act so wildly and expect to maintain employment, even if period readers were impressed by this romantic image of "the hero at work." See Light, "The Ambivalence towards Romance," in *Sinclair Lewis's "Arrowsmith,"* 54.

17. Stuart Chase, *The Tragedy of Waste* (New York: Macmillan, 1925); Cecilia Tichi, *Shifting Gears: Technology, Literature, and Culture in Modernist America* (Chapel Hill: University of North Carolina Press, 1987), 63–75.

18. Alfred Kazin, *On Native Ground* (New York: Doubleday, 1942), 162. Also see Ann Douglas, *Terrible Honesty: Mongrel Manhattan in the 1920s* (New York: Farrar, Straus, and Giroux, 1995).

19. James M. Hutchinsson, *The Rise of Sinclair Lewis: 1920–1930* (University Park, Pa.: Pennsylvania State University Press, 1996), 97; Mark Schorer, *Sinclair Lewis: An American Life* (New York: McGraw-Hill 1961), 337. De Kruif's best-known work is *Microbe Hunters*.

20. Schorer, *Sinclair Lewis: Life*, 337–38. Loeb acquired his belief in the mechanism from the materialistic and reductivist analyses of nineteenth-century German physiologists and from his study of Schopenhauer. See Garland E. Allen, *Life Science in the Twentieth Century* (New York: Wiley, 1975), especially 67, 73ff; and Philip J. Pauly, *Controlling Life: Jacques Loeb and the Engineering Ideal in Biology* (New York: Oxford University Press, 1987), especially 131. De Kruif's esteem for Loeb is evident in his article, "Jacques Loeb, the Mechanist," *Harper's* 146 (January 1923): 182–90.

21. [Paul de Kruif], "Our Medicine-Men: I—Are Commercialism and Science Ruining Medicine?" *Century Magazine* 104 (July 1922): 424–26.

22. [Paul de Kruif], "Medicine," in *Civilization in the United States: An Inquiry by Thirty Americans*, ed. Harold E. Stearns (New York: Harcourt Brace, 1922), 444.

23. [de Kruif], "Our Medicine-Men," 426.

24. Ibid.

25. Ibid., 424. On the importance of bowels in the history of science, see Shapin, "The Philosopher and the Chicken."

26. Lewis to Harcourt et al., 13 February 1923, in *From Main Street to Stockholm*, 125. On the relations between de Kruif and Lewis, see Hutchinsson, *Rise of Sinclair Lewis*, chap. 3; and Richard Lingeman, *Sinclair Lewis: Rebel from Main Street* (New York: Random House, 2002), chap. 15.

27. On the high regard for German education among life scientists, see Thomas Bonner, *American Doctors and German Universities* (Lincoln: University of Nebraska Press, 1963), 107ff.

28. Schorer, "Afterword," 434.

29. Hall, "Research the Vital Spirit," 566.

30. Here, Lewis again echoes de Kruif, who reported that "Science, in its modern definition, is concerned with the quantitative relationship of the factors governing natural phenomena." See [de Kruif], "Medicine," 444.

31. Hutchinsson, *Rise of Sinclair Lewis*, 90.

32. Schorer, "Afterword," 438.

33. Lewis to Grace Lewis, 29 August 1922, in Grace Lewis, *With Love from Gracie: Sinclair Lewis, 1912–1925* (New York: Harcourt Brace, 1955), 212.

34. Hutchinsson, *Rise of Sinclair Lewis*, 95.

35. Kazin notes that Gottlieb was "not merely a European scientist; he was *the* European scientist, the very incarnation of that indescribable cultivation and fathomless European wisdom." Gottlieb certainly bears some attributes of European cultivation but also enacts Lewis's vision of the scientist as barbarian. See Kazin, *On Native Ground*, 177.

36. For a fuller treatment of the St. Hubert section of the book, see Lynch, "*Arrowsmith* Goes Native."

37. Lewis even briefly considered the title "The Gods of Martin Arrowsmith." See Hutchinsson, *Rise of Sinclair Lewis*, 103.

38. Lewis, *From Main Street to Stockholm*, 132.

39. William Morton Wheeler, "The Organization of Research," *Science* 53 (1921): 66. This 1920 speech was cited by anthropologist Robert H. Lowie in the references for his

article "Science," in *Civilization in the United States*, 538, a volume to which de Kruif contributed and probably read.

40. This idea appears to be taken from de Kruif, "Jacques Loeb," 184.

41. Gottlieb describes Terry and Martin as his only "real sons" (effacing his own biological child, Robert Koch Gottlieb [338]). Gottlieb is the authoritative patriarch to Silva's lovable and easygoing "Dad."

EPILOGUE   THE ENDS OF SACRIFICE

1. Herbert W. Horwill, "Scientists at Play," *Scientific American* 98 (4 April 1908): 239; J. O. Perrine, "The Fun of Being a Scientist," *Scientific Monthly* 27 (July 1928): 28–32. Historian Paul Forman argues that after 1940 "fun" supplanted "self-sacrificing humility" as the dominant theme of American physicists' self-images. See his "Social Niche and the Self-Image of the American Physicist," in *The Restructuring of Physical Sciences in Europe and the United States, 1945–1960*, ed. Michelangelo De Maria, Mario Grilli, and Fabio Sebastiani (Singapore: World Scientific Publishing Company, 1989), 96–104. I thank an anonymous reviewer at Rutgers University Press for bringing Forman's claim to my attention.

2. Roselyne Rey, *The History of Pain*, trans. Louise Elliott Wallace, J. A. Cadden, and S. W. Cadden (Cambridge, Mass.: Harvard University Press, 1995); Roy Porter, "Pain and Suffering," in *Companion Encyclopedia of the History of Medicine*, ed. W. F. Bynum and Roy Porter (London: Routledge, 1993), 2:1574–91; *Schloendorff v. Society of New York Hospital*, 106 N.E. 93 (N.Y. 1914); Lederer, *Subjected to Science*; Altman, *Who Goes First?*; Jay Katz, compiler, *Experimentation with Human Beings: The Authority of the Investigator, Subject, Professions, and State in the Human Experimentation Process* (New York: Sage Foundation, 1972); Edward J. Larson, *Summer for the Gods: The Scopes Trial and America's Continuing Debate over Science and Religion* (New York: Basic Books, 1997).

3. See, for instance, "Martyrs of Science," *Literary Digest* 97 (2 June 1928): 19–20; and "Martyrs of Science," *Popular Mechanics* 44 (July 1925): 51.

4. Macpherson, *Political Theory of Possessive Individualism*, 271–72; Mizruchi, *Science of Sacrifice*, chap. 1; Lorraine Daston, *Classical Probability in the Enlightenment* (Princeton, N.J.: Princeton University Press, 1988), chap. 2.

5. Simmel, "Chapter in the Philosophy of Value," 597. See also Mizruchi, *Science of Sacrifice*, 30–31.

6. Georg Simmel, "Exchange," in *On Individuality and Social Forms: Selected Writings*, ed. Donald N. Levine (Chicago: University of Chicago Press, 1971), 52.

7. Baillie, "Self-Sacrifice," 260.

8. Ibid., 261. In its defiance of instrumental calculation, the ethos of sacrifice for science resonated with the ideals of romantic love gaining strength after the Civil War. As Nancy F. Cott explains, increasing emphasis on free choice as the defining principle of American marriage only intensified public discussion of the mysteriously inescapable bonds of love. Reference to the power of uncontrollable, irrational attraction served to accentuate the contrast between arranged or overtly mercenary unions (which evoked "un-American" forms of coercion) and "voluntary" marriages inspired solely by the force of true love. For further discussion of love in the age of contract, see Cott, *Public Vows: A History of Marriage and the Nation* (Cambridge, Mass.: Harvard University Press, 2000), esp. 150–151.

9. Lightman, "Spellbound," D4.

10. Malcolm W. Browne, "Bah! Who Says Science Must Seem Like Fun?" *New York Times*, 13 May 1997, p. C7.

BIBLIOGRAPHY

ARCHIVAL MATERIALS

American Heritage Center, University of Wyoming, Laramie
    Edwin Emory Slosson Collection
Francis A. Countway Library of Medicine, Boston
    Manuscripts and letters regarding Walter J. Dodd
Milton S. Eisenhower Library, Johns Hopkins University, Baltimore
    Daniel Coit Gilman Papers
    Ira Remsen Papers
    Henry Augustus Rowland Papers
Goddard Library, Clark University, Worcester, Massachusetts
    G. Stanley Hall Papers
Hawthorne-Longfellow Library, Bowdoin College, Brunswick, Maine
    Donald Baxter MacMillan Collection
Maine Women Writers Collection, University of New England, Portland, Maine
    Josephine Diebitsch Peary Collection
Manuscript Division, Library of Congress, Washington, D.C.
    Frederick Albert Cook Papers
National Archives and Records Administration, College Park, Maryland
    Rear Admiral Robert E. Peary Family Collection

OTHER SOURCE MATERIAL

Abir-Am, Pnina G., and Dorinda Outram, eds. *Uneasy Careers and Intimate Lives: Women in Science, 1789–1979.* New Brunswick, N.J.: Rutgers University Press, 1987.

Abrahams, Harold J., and Marion B. Savin, eds. *Selections from the Scientific Correspondence of Elihu Thomson.* Cambridge: MIT Press, 1971.

Adams, Cyrus C. "The North Pole at Last." *American Review of Reviews* 40 (1909): 420–26.

Adams, Henry. *The Education of Henry Adams.* New York: Vintage, 1990 [1907].

Adams, James Eli. *Dandies and Desert Saints: Styles of Victorian Masculinity.* Ithaca, N.Y.: Cornell University Press, 1995.

Aeschylus. *Oresteia: Agamemnon, The Libation Bearers, The Eumenides,* translated by Richmond Lattimore. Chicago: University of Chicago Press, 1953.

———. *The Oresteia.* Translated by Hugh Lloyd-Jones. Berkeley: University of California Press, 1993.

Agramonte, Aristides. "The Inside History of a Great Medical Discovery." *Scientific Monthly* 1 (1915): 209–37.

———. "Report of Bacteriological Investigations upon Yellow Fever." *Medical News* 76 (10 February 1900): 203–12.

Ahlstrom, Sydney E. *A Religious History of the American People*. New Haven, Conn.: Yale University Press, 1972.

Allen, Garland E. *Life Science in the Twentieth Century*. New York: Wiley, 1975.

Altman, Lawrence K. *Who Goes First? The Story of Self-Experimentation in Medicine*. New York: Random House, 1987.

Anderson, Amanda. *The Powers of Distance: Cosmopolitanism and the Cultivation of Detachment*. Princeton, N.J.: Princeton University Press, 2001.

Anderson, Bendict. *Imagined Communities: Reflections on the Origin and Spread of Nationalism*. London: Verso, 1983.

Anderson, James D. "The Hampton Model of Normal School Industrial Education, 1868–1900." In *New Perspectives on Black Educational History*, edited by Vincent P. Franklin and James D. Anderson, 61–96. Boston: Hall, 1978.

Anderson, Warwick. "The Trespass Speaks: White Masculinity and Colonial Breakdown." *American Historical Review* 102 (December 1997): 1343–70.

Anthony, William A. "Address." *Proceedings of the American Association for the Advancement of Science* 36 (1887): 69–78.

Arnold, Nick. *Suffering Scientists*. London: Scholastic, 2000.

Arns, Robert G. "The High-Vacuum X-Ray Tube: Technological Change in Social Context." *Technology and Culture* 38 (October 1997): 852–90.

Asad, Talal. "Notes on Body Pain and Truth in Medieval Christian Ritual." *Economy and Society* 12 (1983): 287–327.

Atkinson, Phillip. *Electricity for Everybody; Its Nature and Uses Explained*. 2d ed. New York: Century, 1897.

Baillie, J. B. "Self-Sacrifice," *Hibbert Journal* 12 (January 1914): 260–82.

Baird, Davis. *Thing Knowledge: A Philosophy of Scientific Instruments*. Berkeley: University of California Press, 2004.

Baker, F.W.G. "The First International Polar Year, 1882–83." *Polar Record* 21 (1982): 275–85.

Barad, Karen. "Getting Real: Technoscientific Practices and the Materialization of Reality." *differences* 10, no. 2 (1998): 87–128.

Barker, Elsa. "The Frozen Grail." In *The Little Book of Modern Verse*, edited by Jessie B. Rittenhouse, 119–20. Boston: Houghton Mifflin, [1913].

Barr, William. "The Expeditions of the First International Polar Year, 1882–83." In *The Arctic Institute of North America*. Technical paper 29. Calgary, Alberta: Arctic Institute of North America, 1985.

Baskerville, Charles. "Science and the People." *Science* 20 (26 August 1904): 266–73.

Bataille, Georges. *The Accursed Share: An Essay on General Economy*. Vol. 1. Translated by Robert Hurley. New York: Zone, 1991.

———. *Visions of Excess: Selected Writings, 1927–1939*, edited and translated by Allan Stoekl. Minneapolis: University of Minnesota Press, 1985.

Batnitzky, Leora. "On the Suffering of God's Chosen: Christian Views in Jewish Terms." In *Christianity in Jewish Terms*, edited by Tikva Frymer-Kensky et al., 203–20. Boulder, Colo.: Westview, 2000.

Baym, Nina. *American Women of Letters and the Nineteenth-Century Sciences: Styles of Affiliation*. New Brunswick, N.J.: Rutgers University Press, 2002.

Bean, William B. "Walter Reed and the Ordeal of Human Experiments." *Bulletin of the History of Medicine* 51 (1977): 75–92.

Beattie, Owen, and John Geiger. *Frozen in Time: Unlocking the Secrets of the Franklin Expedition.* New York: Dutton, 1988.

Beck, Carl. *Röntgen Ray Diagnosis and Therapy.* New York: Appleton, 1904.

Becker, Carl L. *The Heavenly City of the Eighteenth-Century Philosophers.* New Haven, Conn.: Yale University Press, 1932.

Bederman, Gail. *Manliness and Civilization: A Cultural History of Gender and Race in the United States, 1880–1917.* Chicago: University of Chicago Press, 1995.

Bellamy, Edward. *Doctor Heidenhoff's Process.* New York: AMS Press, 1969 [1880].

Bending, Lucy. *The Representation of Bodily Pain in Late Nineteenth-Century English Culture.* Oxford: Oxford University Press, 2000.

Bergonié, J., L. Tribondeau, and D. Récamier. "Action des rayons X sur l'ovaire de la lapine." *Comptes rendus hebdomadaires des séances et mémoires de la société de biologie* 58 (17 February 1905): 284–86.

Berlant, Lauren. "National Brands/National Body: Imitation of Life." In *Comparative American Identities: Race, Sex, and Nationality in the Modern Text,* edited by Hortense J. Spillers, 110–40. New York: Routledge, 1991.

Berton, Pierre. *The Arctic Grail: The Quest for the Northwest Passage and the North Pole, 1818–1909.* New York: Viking, 1988.

Bhabha, Homi K. "DissemiNation: Time, Narrative, and the Margins of the Modern Nation." In *Nation and Narration,* edited by Homi K. Bhabha, 291–322. London: Routledge, 1990.

Biagioli, Mario. "Tacit Knowledge, Courtliness, and the Scientist's Body." In *Choreographing History,* edited by Susan Leigh Foster, 69–81. Bloomington: Indiana University Press, 1995.

Bigelow, Henry Jacob. *Surgical Anaesthesia: Addresses and Other Papers.* Boston: Little, Brown, 1900.

Birkett, Dea. *Spinsters Abroad: Victorian Lady Explorers.* New York: Blackwell, 1989.

Bishop, Sande. *Radiology in New England: The First Hundred Years (1896–1995).* Boston: Massachusetts Radiological Society and New England Roentgen Ray Society, 1995.

Bleich, Alan Ralph. *The Story of X-Rays from Röntgen to Isotopes.* New York: Dover, 1960.

Blight, David W. *Race and Reunion: The Civil War in American Memory.* Cambridge, Mass.: Harvard University Press, 2001.

Bloom, Harold, ed. *Sinclair Lewis's "Arrowsmith."* New York: Chelsea House, 1988.

Bloom, Lisa. *Gender on Ice: American Ideologies of Polar Expeditions.* Minneapolis: University of Minnesota Press, 1993.

Bonner, Thomas. *American Doctors and German Universities.* Lincoln: University of Nebraska Press, 1963.

Borden, W. C. *The Use of the Röntgen Ray by the Medical Department of the United States Army in the War with Spain 1898.* Washington, D.C.: U.S. Government Printing Office, 1900.

Boscagli, Maurizia. *Eye on the Flesh: Fashions of Masculinity in the Early Twentieth Century.* Boulder, Colo.: Westview, 1996.

Bourdieu, Pierre. *Outline of a Theory of Practice,* translated by Richard Nice. Cambridge, U.K.: Cambridge University Press, 1977.

Bourke, Joanna. *Dismembering the Male: Men's Bodies, Britain, and the Great War.* Chicago: University of Chicago Press, 1996.

Boyarin, Daniel. *Dying for God: Martyrdom and the Making of Christianity and Judaism.* Stanford, Calif.: Stanford University Press, 1999.

Bradford, William. *The Arctic Regions: Illustrated with Photographs Taken on an Art Expedition to Greenland.* London: Sampson Low, Marston, Low, and Searle, 1873.

Bragg, Rick. "A Drop of Monkey Virus Kills a Researcher in 6 Weeks." *New York Times*, 14 December 1997, p. 29.

Bravo, Michael T. "The Accuracy of Ethnoscience: A Study of Inuit Cartography and Cross-Cultural Commensurability." *Manchester Papers in Social Anthropology* 2 (1996): 1–36.

Brecher, Ruth, and Edward Brecher. *The Rays: A History of Radiology in the United States and Canada*. Baltimore: Williams and Wilkins, 1969.

Brewster, Sir David. *The Martyrs of Science*. New York: Harper, 1841.

Brickner, Samuel L. "On the Physiological Character of the Pain of Parturition." *Gaillard's Medical Journal* 72 (January 1900): 794–801.

Briggs, Charles A. "The Salvation Army." *North American Review* 159 (1894): 697–710.

Briggs, Laura. "The Race of Hysteria: 'Overcivilization' and the 'Savage' Woman in Late Nineteenth-Century Obstetrics and Gynecology." *American Quarterly* 52 (June 2000): 246–73.

Brown, Norman O. *Apocalypse and/or Metamorphosis*. Berkeley: University of California Press, 1991.

——. *Love's Body*. Berkeley: University of California Press, 1966.

Brown, Percy. *American Martyrs to Science through the Roentgen Rays*. Springfield, Ill.: Thomas, 1936.

——. "Walter J. Dodd." *American Journal of Roentgenology* 4 (1917): 629–30.

Brown, Wendy. *States of Injury: Power and Freedom in Late Modernity*. Princeton, N.J.: Princeton University Press, 1995.

Brown-Séquard, Charles-Edouard. "Des Effets produits chez l'homme par des injections sous-cutanées une liquide retiré des testicules frais de cobaye et de chien." *Comptes rendus hebdomadaires de séances et mémoires de la Sociéte de Biologie* 1 (1889): 415–19.

Browne, Malcolm W. "Bah! Who Says Science Must Seem Like Fun?" *New York Times*, 13 May 1997, p. C-7.

Bruce, Robert V. *The Launching of Modern American Science, 1846–1876*. Ithaca, N.Y.: Cornell University Press, 1987.

Bryce, Robert. *Cook & Peary: The Polar Controversy, Resolved*. Mechanicsburg, Pa.: Stackpole, 1997.

Burnett, D. Graham. *Masters of All They Surveyed: Exploration, Geography, and a British El Dorado*. Chicago: University of Chicago Press, 2000.

Butler, Judith. *Gender Trouble: Feminism and the Subversion of Identity*. New York: Routledge, 1990.

——. *Undoing Gender*. New York: Routledge, 2004.

Cantor, Geoffrey. "The Scientist As Hero: Public Images of Michael Faraday." In *Telling Lives in Science: Essays on Scientific Biography*, edited by Michael Shortland and Richard Yeo, 171–93. Cambridge, U.K.: Cambridge University Press, 1996.

Carhart, Henry S. "The Educational and Industrial Value of Science." *Science* 1 (12 April 1895): 393–402.

——. "The Humanistic Element in Science." *Science* 4 (31 July 1896): 124–30.

Carnegie, Charles V. "The Dundus and the Nation." *Cultural Anthropology* 11, no. 4 (1996): 470–509.

Carnes, Mark C. *Secret Ritual and Manhood in Victorian America*. New Haven, Conn.: Yale University Press, 1989.

Carnes, Mark C., and Clyde Griffen, eds. *Meanings for Manhood: Constructions of Masculinity in Victorian America*. Chicago: University of Chicago, 1990.

Cartwright, Lisa. *Screening the Body: Tracing Medicine's Visual Culture*. Minneapolis: University of Minnesota, 1995.

Chase, Stuart. *The Tragedy of Waste*. New York: Macmillan, 1925.

Chew, Lee. "The Life Story of a Chinaman." In *The Life Stories of Undistinguished Americans, As Told by Themselves*, edited by Hamilton Holt, 174–85. New York: Routledge, 1990 [1903].

Christie, Arthur C. "Fifty Years of Progress in Radiology." In *The American Roentgen Ray Society, 1900–1950: Commemorating the Golden Anniversary of the Society*, 25–29. Springfield, Ill: Thomas, 1950.

Christie, J.R.R. "The Paracelsian Body." In *Paracelsus: The Man and His Reputation, His Ideas and Their Transformation*, edited by Ole Peter Grell, 269–91. Leiden: Brill, 1998.

Clark, Elizabeth B. " 'The Sacred Rights of the Weak': Pain, Sympathy, and the Origins of Humanitarian Sensibility." *Journal of American History* 82, no. 2 (1995): 463–93.

Clarke, F. W. "American Colleges versus American Science." *Popular Science Monthly* 9 (1876): 467–79.

Clemenceau, Georges. *American Reconstruction, 1865–1870*, edited by Fernand Baldensperger, translated by Margaret MacVeagh. New York: Da Capo, 1969.

Codman, E. A. "The Cause of Burns from X-Rays." *Boston Medical and Surgical Journal* 135 (10 December 1896): 610–11.

Cohen, Esther. "The Animated Pain of the Body." *American Historical Review* 105 (February 2000): 36–68.

Cohen, I. Bernard, ed. *Puritanism and the Rise of Modern Science: The Merton Thesis*. New Brunswick, N.J.: Rutgers University Press, 1990.

Cole, Lewis Gregory. "In Memorium." *American Journal of Roentgenology* 4 (1917): 629–30.

Collier, Stephen J., and Andrew Lakoff. "On Regimes of Living." In *Global Assemblages: Technology, Politics, and Ethics As Anthropological Problems*, edited by Aihwa Ong and Stephen J. Collier, 22–39. New York: Blackwell, 2005.

Colwell, Hector Alfred, and Sidney Russ. *X-Ray and Radium Injuries, Prevention and Treatment*. London: Oxford University Press, 1934.

"Comment and Criticism." *Science* 3 (29 February 1884): 241.

"Comment and Criticism." *Science* 3 (2 May 1884): 530.

Cook, Frederick A. *My Attainment of the Pole: Being the Record of the Expedition that First Reached the Boreal Center, 1907–1909, With the Final Summary of the Polar Controversy*. New York: Kennerley, 1912.

Cooke, Josiah P., Jr. "Scientific Culture." *Popular Science Monthly* 7 (1875): 513–31.

Cordasco, Francesco. *Daniel Coit Gilman and the Protean Ph.D: The Shaping of American Graduate Education*. Leiden: Brill, 1960.

Cott, Nancy F. *Public Vows: A History of Marriage and the Nation*. Cambridge, Mass.: Harvard University Press, 2000.

Coulter, John M. *Mission of Science in Education*. Ann Arbor: University of Michigan, 1900.

Counter, S. Allen. *North Pole Legacy: Black, White and Eskimo*. Amherst: University of Massachusetts Press, 1991.

Creighton, Margaret S., and Lisa Norling, eds. *Iron Men, Wooden Women: Gender and Seafaring in the Atlantic World, 1700–1920*. Baltimore: Johns Hopkins University Press, 1996.

Croce, Paul Jerome. "A Useful Eccentricity: William James's Engagement with Science." *Isis* 93 (2002): 272–76.

Cunningham, Andrew. "Getting the Game Right: Some Plain Words on the Identity and Invention of Science." *Studies in the History and Philosophy of Science* 19 (1988): 365–89.

Curtis, Heber D. "Navigation Near the Pole." *United States Naval Institute Proceedings* 65 (January 1939): 9–19.

Daniel, John. "The X-Rays." *Science* 3 (10 April 1896): 562–63.

Daniels, George H., ed. *Nineteenth-Century American Science: A Reappraisal.* Evanston: North-western University Press, 1972.

———. "The Pure-Science Ideal and Democratic Culture." *Science* 156 (30 June 1967): 1699–1705.

Darwin, Charles. *The Expression of the Emotions in Man and Animals.* New York: Greenwood, 1955 [1872].

Daston, Lorraine. "Academies and the Unity of Knowledge: The Disciplining of the Disciplines." *differences* 10, no. 2 (1998): 67–86.

———. "Before Vocation: Science As Work." Paper presented at the University of California, Los Angeles, 17 May 2003.

———. *Classical Probability in the Enlightenment.* Princeton, N.J.: Princeton University Press, 1988.

———. "Fear and Loathing of the Imagination in Science." *Daedalus* 127 (winter 1997): 73–95.

———. "Objectivity and the Escape from Perspective." *Social Studies of Science* 22 (1992): 597–618.

———. "Objectivity and the Scientific Self." Paper presented at Rutgers University, New Brunswick, N.J., 24 February 2003.

Daston, Lorraine, and Peter Galison. "The Image of Objectivity." *Representations* 40 (1992): 81–128.

Davis, Angela Y. "From the Prison of Slavery to the Slavery of Prison: Frederick Douglass and the Convict Lease System." In *Frederick Douglass: A Critical Reader*, edited by Bill E. Lawson and Frank M. Kirkland, 339–62. Malden, Mass.: Blackwell, 1999.

Davis, C. H., ed. *Narrative of the North Polar Expedition: U.S. Ship Polaris, Captain Charles Francis Hall Commanding.* Washington, D.C.: U.S. Government Printing Office, 1876.

Day, J. P. "Locke on Property." *Philosophical Quarterly* 16 (July 1966): 207–20.

Dear, Peter. "From Truth to Disinterestedness in the Seventeenth Century." *Social Studies of Science* 22 (1992): 619–31.

"Death of a Famous Woman Radiographer," *San Francisco Chronicle*, 5 August 1905, p. 10.

Decker, Karl. "Dr. Frederick A. Cook—Faker." *Metropolitan* 31 (January 1910): 416–35.

de Kruif, Paul. "Jacques Loeb, the Mechanist." *Harper's* 146 (January 1923): 182–90.

[———]. "Medicine." In *Civilization in the United States: An Inquiry by Thirty Americans*, edited by Harold E. Stearns, 443–56. New York: Harcourt Brace, 1922.

[———]. "Our Medicine-Men: I—Are Commercialism and Science Ruining Medicine?" *Century Magazine* 104 (July 1922): 416–26.

———. *Microbe Hunters.* New York: Pocket Books, 1926.

———. *The Sweeping Wind: A Memoir.* New York: Harcourt, Brace, and World, 1962.

Delaporte, François. *The History of Yellow Fever: An Essay on the Birth of Tropical Medicine*, translated by Arthur Goldhammer. Cambridge, Mass.: MIT Press, 1991.

"Deleterious Effects of X Rays on the Human Body." *Electrical Review* 29 (12 August 1896): 78.

De Long, Emma, ed. *The Voyage of the Jeannette: The Ship and Ice Journals of George W. De Long, Lieutenant-Commander U.S.N., and Commander of the Polar Expedition of 1879–1881.* Vol. 2. Boston: Houghton Mifflin, 1884.

de Moulin, Daniel. "A Historical-Phenomenological Study of Bodily Pain in Western Man." *Bulletin of the History of Medicine* 48 (1984): 540–70.

Derrida, Jacques. *Given Time: I. Counterfeit Money*, translated by Peggy Kamuf. Chicago: University of Chicago Press, 1992.

Desjardins, Arthur U. "The Status of Radiology in America." *Journal of the American Medical Association* 92 (30 March 1929): 1035–39.

Dewey, John. *The Influence of Darwinism on Philosophy and Other Essays in Contemporary Thought*. Bloomington: Indiana University Press, 1965 [1910].

Dewing, Stephen B. *Modern Radiology in Historical Perspective*. Springfield, IL: Thomas, 1962.

Dick, Steven. "Simon Newcomb, William Harkness, and the Nineteenth-Century American Transit of Venus Expedition." *Journal for the History of Astronomy* 29, no. 3 (1998): 221–55.

Dillard, Annie. "An Expedition to the Pole." In *Surviving Crisis*, edited by Lee Gutkind, 80–105. New York: Tarcher/Putnam, 1997.

Donizetti, Pino. *Shadow and Substance: The Story of Medical Radiography*, translated by Anne Ellis. Oxford: Pergamon, 1967.

Douglas, Ann. *Terrible Honesty: Mongrel Manhattan in the 1920s*. New York: Farrar, Straus, and Giroux, 1995.

duBois, Page. *Torture and Truth*. New York: Routledge, 1991.

Duffin, Jacalyn, and Charles R. R. Hayter. "Baring the Sole: The Rise and Fall of the Shoe-Fitting Fluoroscope." *Isis* 91 (2000): 260–82.

Durkheim, Emile. *The Elementary Forms of Religious Life*, translated by Joseph Ward Swain. London: Allen and Unwin, 1915.

Early Proceedings of the American Philosophical Society. Philadelphia: McCalla and Stavely, 1884.

"The Edison Fluoroscope Exhibit," *Electrical Engineer* 21 (3 June 1896): 600, 601.

"Edison and X-Ray Injuries." *Journal of the American Medical Association* 41 (22 August 1903): 499.

"Editor's Table: Purpose and Plan of Our Enterprise." *Popular Science Monthly* 1 (May 1872): 113–15.

Edwards, Jonathan. *Representative Selections*, edited by Clarence H. Faust and Thomas H. Johnson. New York: Hill and Wang, 1962.

———. *The Works of President Edwards*. Vol. 4. New York: Leavitt and Allen, 1851.

Eisenberg, Ronald L. *Radiology: An Illustrated History*. St. Louis, Mo.: Mosby, 1992.

Elazar, Daniel J. ed.. *Covenant in the Nineteenth Century: The Decline of an American Tradition* Lanham, Md.: Rowman and Littlefield, 1994.

Eliot, T. S. "Tradition and Individual Talent." In *The Sacred Wood: Essays on Poetry and Criticism*, 47–59. London: Methuen, 1948 [1920].

Elliott, Clark A. *History of Science in the United States: A Chronology and Research Guide*. New York: Garland, 1996.

———. "Scientists, Nature and the Self: Reading Autobiographical Writings from Nineteenth Century America." Paper presented at the annual meeting of the History of Science Society, San Diego, 1997.

Elliott, Michael A. "Telling the Difference: Nineteenth-Century Narratives of Racial Taxonomy." *Law and Social Inquiry* 24, no. 3 (1999): 611–36.

Emerson, Ralph Waldo. *The Complete Works of Ralph Waldo Emerson*. Vol. 9. Boston: Houghton Mifflin, 1904.

———. *Selected Writings of Emerson*, edited by Donald McQuade. New York: Random House, 1981.

Faust, Drew Gilpin. "Altars of Sacrifice: Confederate Women and the Narratives of War." *Journal of American History* 76 (March 1990): 1200–28.

Feuer, Lewis S. *The Scientific Intellectual: The Psychological and Sociological Origins of Modern Science*. New York: Basic Books, 1963.

Ffirth, Stubbins. *A Treatise on Malignant Fever: With an Attempt to Prove Its Non-Contagious Nature*. Philadelphia: B. Graves, 1804.

Fishbein, Morris. *The New Medical Follies; An Encyclopedia of Cultism and Quackery in these United States*. New York: Boni and Liveright, 1927.

Fogg, G. E. *A History of Antarctic Science*. New York: Cambridge University Press, 1992.

Foner, Eric. "The Meaning of Freedom in the Age of Emancipation." *Journal of American History* 81 (September 1994): 435–60.

———. *The Story of American Freedom*. New York: Norton, 1998.

Forgan, Sophie. "The Architecture of Science and the Idea of a University." *Studies in the History and Philosophy of Science* 20, no. 4 (1989): 405–34.

Forman, Paul. "Social Niche and the Self-Image of the American Physicist." In *The Restructuring of Physical Sciences in Europe and the United States, 1945–1960*, edited by Michelangelo De Maria, Mario Grilli, and Fabio Sebastiani, 96–115. Singapore: World Scientific Publishing Company, 1989.

Foucault, Michel. "Body/Power." In *Power/Knowledge: Selected Interviews and Other Writings*, edited by Colin Gordon, translated by Colin Gordon et al., 55–62. New York: Pantheon, 1980.

———. *Discipline and Punish*, translated by A. M. Sheridan. London: Lande, 1977 [1975].

———. *Language, Counter-Memory, Practice: Selected Essays and Interviews*, edited by Donald F. Bouchard, translated by Donald F. Bouchard and Sherry Simon. Ithaca, N.Y.: Cornell University Press, 1977.

———. *Ethics: Subjectivity and Truth; Essential Works of Foucault, 1954–1984*, edited by Paul Rabinow, translated by Robert Hurley et al. New York: New Press, 1997.

———. *The Use of Pleasure*. Vol. 2 of *The History of Sexuality*, translated by Robert Hurley. New York: Vintage, 1990.

Foxe, John. *Foxe's Book of Martyrs*. Springdale, Pa.: Whitaker House, 1981 [1563].

Frängsmyr, Tore, ed. *Solomon's House Revisited: The Organization and Institutionalization of Science*. Canton, Mass.: Science History Publications, 1990.

Franklin, John Hope, and Alfred A. Moss, Jr. *From Slavery to Freedom: A History of African Americans*. 8th ed. New York: Knopf, 2000.

Frazer, Sir James George. *The Golden Bough: A Study in Magic and Religion*. New York: Macmillan, 1922 [1890].

Frederickson, George M. *Inner Civil War: Northern Intellectuals and the Crisis of the Union*. New York: Harper and Row, 1965.

Freedgood, Elaine. *Victorian Writing about Risk: Imagining a Safe England in a Dangerous World*. Cambridge, U.K.: Cambridge University Press, 2000.

Frei, G. A. "Deleterious Effects of X Rays on the Human Body: Further Evidence that Repeated Exposure to the Rays Produces a Sunburn Effect." *Electrical Review* 29 (19 August 1896): 95.

———. "X-Rays Harmless with the Static Machine." *Electrical Engineer* 22 (23 December 1896): 651.

———. "X-Rays Harmless with the Static Machine." *Electrical World* 29 (2 January 1897): 27.

———. "X-Ray Physiological Effects." *Electrical Engineer* 22 (16 September 1896): 276.

Freud, Sigmund. "On the Universal Tendency to Debasement in the Sphere of Love." In *The Standard Edition to the Complete Works of Sigmund Freud*, translated by James Strachey. Vol. 11, pp. 178–90. London: Hogarth, 1957 [1912].

Fuchs, W. C. "Effect of Röntgen Rays on the Skin." *Western Electrician* 19 (12 December 1896): 291.

"The Future of American Science." *Science* 1 (1883): 1–3.

Gage, W. L. *Introduction to Our Lost Explorers: The Narrative of the Jeannette Arctic Expedition, as Related by the Survivors, and in the Records and Last Journals of Lieutenant De Long,* edited by Raymond Lee Newcomb. Hartford, Conn: American Publishing Company, 1882.

Galison, Peter. "Judgement Against Objectivity." In *Picturing Science, Producing Art*, edited by Peter Galison and Caroline Jones, 327–59. New York: Routledge, 1998.

Galison, Peter, and David J. Stump, ed. *The Disunity of Science: Boundaries, Contexts, and Power.* Stanford, Calif.: Stanford University Press, 1996.

Getman, Frederick H. *The Life of Ira Remsen.* Easton, Pa.: Journal of Chemical Education, 1940.

Giedion, Siegfried. *Mechanization Takes Command: A Contribution to Anonymous History.* New York: Oxford University Press, 1948.

Gilman, Daniel Coit. *Addresses at the Inauguration of Daniel C. Gilman as President of the Johns Hopkins University.* Baltimore: Murphy, 1876.

———. *The Building of the University: An Inaugural Address Delivered at Oakland, November 7th, 1872.* San Francisco: Carmany, 1872.

Girard, René. *Violence and the Sacred.* Baltimore: Johns Hopkins University Press, 1977.

Glare, P.G.W., ed. *Oxford Latin Dictionary.* New York: Oxford University Press, 1997.

Glasser, Otto. *Wilhelm Conrad Röntgen and the Early History of Röntgen Rays.* Springfield, Ill.: Thomas, 1934.

Glenn, Myra C. *Campaigns against Corporal Punishment: Prisoners, Sailors, Women, and Children in Antebellum America.* Albany: State University of New York Press, 1984.

Gobineau, Arthur de. *The Moral and Intellectual Diversity of the Races.* Philadelphia: Lippincott, 1856.

Goetzmann, William H. "A 'Capacity for Wonder': The Meanings of Exploration." In *North American Exploration*, edited by John Logan Allen, vol. 3, pp. 521–45. Lincoln: University of Nebraska Press, 1997.

Golan, Tal. *Laws of Men and Laws of Nature: The History of Scientific Expert Testimony in England and America.* Cambridge, Mass.: Harvard University Press, 2004.

Goldberg, Carey. "Colleagues Vow to Learn from Chemist's Death." *New York Times*, 3 October 1997, p. A7.

Golinski, Jan. "The Care of the Self and the Masculine Birth of Science." *History of Science* 40 (2002): 125–45.

Goliszek, Andrew. *In the Name of Science: A History of Secret Programs, Medical Research, and Human Experimentation.* New York: St. Martin's, 2003.

Good, Mary-Jo DelVecchio, Paul E. Brodwin, Byron J. Good, and Arthur Kleinman, eds. *Pain As Human Experience: An Anthropological Perspective.* Berkeley: University of California Press, 1992.

Goode, G. Brown. "America's Relation to the Advance of Science." *Science* 1 (4 January 1895): 4–9.

Goodman, Jordan, Anthony McElligott, and Lara Marks, eds. *Useful Bodies: Humans in the Service of Medical Science in the Twentieth Century.* Baltimore: Johns Hopkins University Press, 2003.

Goonatilake, Susantha. "Modern Science and the Periphery: The Characteristics of Dependent Knowledge." In *The "Racial" Economy of Science: Toward a Democratic Future*, edited by Sandra Harding, 259–67. Bloomington: Indiana University Press, 1993.

Greely, Adolphus Washington. *Three Years of Arctic Service: An Account of the Lady Franklin Bay Expedition of 1881–84.* 2 vols. New York: Scribner, 1886.

Grell, Ole Peter, and Andrew Cunningham, eds. *Religio Medici: Medicine and Religion in Seventeenth-Century England.* Aldershot, U.K.: Scolar, 1996.

Griffith, R. Marie. "Apostles of Abstinence: Fasting and Masculinity during the Progressive Era." *American Quarterly* 52 (December 2000): 599–638.

Grigg, E.R.N. *The Trail of the Invisible Light.* Springfield, Ill.: Thomas, 1965.

"Group's Members Say They'll Risk Lives in AIDS Vaccine Tests." *Los Angeles Times*, 22 September 1997, p. AII.

Gross, Charles. "Address." In *Williams College: Centennial Anniversary, 1793–1893*, 173–203. Cambridge, Mass.: Wilson, 1894.

Grosz, Elizabeth. *Volatile Bodies: Toward a Corporeal Feminism.* Bloomington: Indiana University Press, 1994.

Grubbé, Emil H. *X-Ray Treatment: Its Origin, Birth, and Early History.* Saint Paul, Minn.: Bruce, 1949.

Guillory, James Denny. "The Pro-Slavery Arguments of Dr. Samuel A. Cartwright." *Louisiana History* 9 (1968): 209–27.

Guralnick, Stanley M. *Science and the Ante-Bellum American College.* Philadelphia: American Philosophical Society, 1975.

Gutman, Herbert G. *Work, Culture, and Society in Industrializing America.* New York: Knopf, 1977.

Habermas, Jürgen. "Modernity: An Incomplete Project." In *The Anti-Aesthetic*, edited by Hal Foster, 3–15. Seattle: Bay, 1983.

Hagstrom, Warren O. "Gift Giving As an Organizing Principle in Science." In *Science in Context: Readings in the Sociology of Science*, edited by Barry Barnes and David Edge, 21–34. Cambridge, Mass.: MIT Press, 1982.

Hall, G. Stanley. "Confessions of a Psychologist, Part I." *Pedagogical Seminary* 8 (1901): 92–143.

———. *Life and Confessions of a Psychologist.* New York: Appleton, 1923.

———. "New Departures in Education." *North American Review* 140 (February 1885): 144–52.

———. "Overpressure in Schools." *Nation* 41 (22 October 1885): 338–39.

———. "Research the Vital Spirit of Teaching." *Forum* 17 (1894): 558–70.

———. "The University Idea." *Pedagogical Seminary* 15 (1908): 92–104.

Halsted, George Bruce. "The Culture Given by Science." *Science* 4 (3 July 1896): 12–13.

Halttunen, Karen. *Confidence Men and Painted Women: A Study of Middle-Class Culture in America.* New Haven, Conn.: Yale University Press, 1982.

———. "Humanitarianism and the Pornography of Pain in Anglo-American Culture." *American Historical Review* 100 (April 1995): 303–34.

Hamilton, Gail. "A Call to My Country-Women." *Atlantic Monthly* 11 (March 1863): 345–49.

Hannaway, Owen. "The German Model of Chemical Education in America: Ira Remsen at Johns Hopkins (1876–1913)." *Ambix* 23 (1976): 145–64.

Haraway, Donna J. *Modest_Witness@Second_Millennium.FemaleMan©_Meets_OncoMouse™: Feminism and Technoscience.* New York: Routledge, 1997.

———. *Primate Visions: Gender, Race, and Nature in the World of Modern Science.* New York: Routledge, 1989.

———. *Simians, Cyborgs, and Women: The Reinvention of Nature.* New York: Routledge, 1991.

Harding, Sandra, ed. *The "Racial" Economy of Science: Toward a Democratic Future.* Bloomington: Indiana University Press, 1993.

Harpham, Geoffrey Galt. *The Ascetic Imperative in Culture and Criticism.* Chicago: University of Chicago, 1987.

Harris, Cheryl I. "Whiteness As Property." In *Critical Race Theory: The Key Writings That Formed the Movement*, edited by Kimberlé Crenshaw, Neil Gotanda, Gary Peller, and Kendall Thomas, 276–91. New York: New Press, 1995.

Harrison, Robert Pogue. *Forests: The Shadow of Civilization.* Chicago: University of Chicago Press, 1992.

Hart, James Morgan. *German Universities: A Narrative of Personal Experience*. New York: Putnam, 1874.

"Hart's German Universities." *Nation* 19 (17 December 1874): 400–401.

Hartman, Saidiya V. *Scenes of Subjection: Terror, Slavery, and Self-Making in Nineteenth-Century America*. New York: Oxford University Press, 1997.

Hawkins, Hugh. *Pioneer: A History of the Johns Hopkins University, 1874–1889*. Ithaca, N.Y.: Cornell University Press, 1960.

———. "University Identity: The Teaching and Research Functions." In *The Organization of Knowledge in America, 1860–1920*, edited by Alexandra Oleson and John Voss, 293–94. Baltimore: Johns Hopkins University Press, 1979.

Hawks, H. D. "The Physiological Effects of the Roentgen Rays." *Electrical Engineer* 22 (16 September 1896): 276.

Hawthorne, Nathaniel. *Hawthorne's Short Stories*, edited by Newton Arvin. New York: Vintage, 1946.

Hayles, N. Katherine. *How We Became Posthuman: Virtual Bodies in Cybernetics, Literature, and Informatics*. Chicago: University of Chicago Press, 1999.

Haynes, Roslynn D. *From Faust to Strangelove: Representations of the Scientist in Western Literature*. Baltimore: Johns Hopkins University Press, 1994.

Hegel, G.W.F. *The Encyclopaedia Logic*, translated by Theodore F. Geraets, W. A. Suchting, and H. S. Harris. Indianapolis: Hackett, 1991.

Hendrick, Burton J. *The Training of an American: The Earlier Life and Letters of Walter H. Page, 1855–1913*. Boston: Houghton Mifflin, 1928.

Henson, Matthew A. *A Negro Explorer at the North Pole*. New York: Arno and the New York Times, 1969.

Herndl, Diane Price. "The Invisible (Invalid) Woman: African-American Women, Illness, and Nineteenth-Century Narrative." In *Women and Health in America*, edited by Judith Walzer Leavitt, 2d ed., 131–45. Madison: University of Wisconsin Press, 1999.

Herschbach, Lisa. "Prosthetic Reconstructions: Making the Industry, Re-making the Body, Modelling the Nation," *History Workshop Journal* 44 (1997): 22–57.

Herzig, Rebecca. "On Performance, Productivity, and Vocabularies of Motive in Recent Studies of Science." *Feminist Theory* 5, no. 2 (2004): 127–47.

———. "Removing Roots: 'North American Hiroshima Maiden Syndrome' and the X-Ray." *Technology and Culture* 40 (October 1999): 723–45.

Hesford, Wendy S., and Wendy Kozol, eds. *Haunting Violations: Feminist Criticism and the Crisis of the "Real."* Chicago: University of Illinois Press, 2001.

Hevly, Bruce. "The Heroic Science of Glacier Motion." *Osiris* 11 (1996): 66–86.

Hickey, P. M. "The First Decade of American Roentgenology." *American Journal of Roentgenology* 20 (1928): 150–57.

Hirsch, J. M. "Scientists Stunned by Fatal Mercury Poisoning of Colleague." *San Diego Union-Tribune*, 11 June 1997, p. A11.

Hirschman, Albert O. *The Passions and the Interests: Political Arguments for Capitalism before Its Triumph*. Princeton, N.J.: Princeton University Press, 1977.

Hochman, David. "Buried Alive and Loving It." *National Geographic Adventure* (December 2002–January 2003): 51–57.

Hodges, Paul C. *The Life and Times of Emil H. Grubbé*. Chicago: University of Chicago Press, 1964.

Hofstader, Richard, and C. DeWitt Hardy. *The Development and Scope of Higher Education in the United States*. New York: Columbia University Press, 1952.

Hollinger, David A. "Inquiry and Uplift: Late Nineteenth-Century American Academics and the Moral Efficacy of Scientific Practice." In *The Authority of Experts*, edited by Thomas L. Haskell, 142–56. Bloomington: Indiana University Press, 1984.

Horwill, Herbert W. "Scientists at Play." *Scientific American* 98 (4 April 1908): 239.

Horwitz, Morton J. *The Transformation of American Law, 1870–1960: The Crisis of Legal Orthodoxy*. New York: Oxford University Press, 1992.

Hounshell, David A. "Edison and the Pure Science Ideal in America." *Science* 207 (1980): 612–17.

Howard, Sidney *Yellow Jack*. New York: Harcourt Brace, 1934.

Howell, Joel D. *Technology in the Hospital: Transforming Patient Care in the Early Twentieth Century*. Baltimore: Johns Hopkins University Press, 1995.

Hubert, Henri, and Marcel Mauss. *Sacrifice: Its Nature and Function*, translated by W. D. Halls. Chicago: University of Chicago Press, 1981 [1898].

Hunt, William R. *To Stand at the Pole: The Dr. Cook–Admiral Peary North Pole Controversy*. New York: Stein and Day, 1981.

Hutchinsson, James M. *The Rise of Sinclair Lewis: 1920–1930*. University Park: Pennsylvania State University Press, 1996.

Huxley, Thomas H. *Science and Education: Essays*. New York: Appleton, 1894.

Hyde, Alan. *Bodies of Law*. Princeton, N.J.: Princeton University Press, 1997.

Inda, Jonathan Xavier. "Performativity, Materiality, and the Racial Body." *Latino Studies Journal* II (fall 2000): 74–99.

Irigaray, Luce. *Speculum of the Other Woman*, translated by Gillian C. Gill. Ithaca, N.Y.: Cornell University Press, 1985.

Jackson, Jean. "Chronic Pain and the Tension Between the Body As Subject and Object." In *Embodiment and Experience: The Existential Ground of Culture and Self*, edited by Thomas J. Csordas, 201–28. New York: Cambridge University Press, 1994.

James, William. *The Essential Writings*, edited by Bruce W. Wilshire. Albany: State University of New York Press, 1989.

———. *The Varieties of Religious Experience*. Cambridge, Mass.: Harvard University Press, 1985 [1902].

———. *The Will to Believe and Other Essays in Popular Philosophy*. New York: Longmans, Green, 1919.

Jameson, Fredric. "Postmodernism and Consumer Society." In *The Anti-Aesthetic*, edited by Hal Foster, III–25. Seattle: Bay, 1983.

Jay, James M. *Negroes in Science: Natural Science Doctorates, 1876–1969*. Detroit: Balamp, 1971.

Johnson, Joseph Taber. "On Some of the Apparent Peculiarities of Parturition in the Negro Race, with Remarks on Race Pelves in General." *American Journal of Obstetrics* 8 (January 1875): 88–123.

Johnson, Walter. "The Slave Trader, the White Slave, and the Politics of Racial Determination in the 1850s." *Journal of American History* 87 (June 2000): 13–38.

———. *Soul by Soul: Life Inside the Antebellum Slave Market*. Cambridge, Mass.: Harvard University Press, 1999.

Johnston, William Preston. *The Work of the University in America*. Columbia, S.C.: Presbyterian Publishing House, 1884.

Jones, Fred S. "Deleterious Effects of X Rays on the Human Body." *Electrical Review* 29 (9 September 1896): 127.

Jordanova, Ludmilla. "Presidential Address: Remembrance of Science Past." *British Journal for the History of Science* 33 (2000): 387–406.

Kantorowicz, Ernst H. *The King's Two Bodies: A Study in Mediaeval Political Theology.* Princeton, N.J.: Princeton University Press, 1997 [1957].

Kaplan, Amy, and Donald E. Pease, eds. *Cultures of United States Imperialism.* Durham, N.C.: Duke University Press, 1993.

Kassabian, Mihran K. *Roentgen Rays and Electro-Therapeutics.* Philadelphia: Lippincott, 1907.

———. "X-Ray as an Irritant." *American X-Ray Journal* 7 (1900): 784–86.

Kasson, John. *Rudeness and Civility: Manners in Nineteenth-Century Urban America.* New York: Hill and Wang, 1990.

Katz, Jay, compiler. *Experimentation with Human Beings: The Authority of the Investigator, Subject, Professions, and State in the Human Experimentation Process.* New York: Sage Foundation, 1972.

Kazin, Alfred. *On Native Ground.* New York: Doubleday, 1942.

Keller, Evelyn Fox. *A Feeling for the Organism: The Life and Work of Barbara McClintock.* New York: Freeman, 1983.

———. "The Paradox of Scientific Subjectivity." *Annals of Scholarship* 9 (1992): 135–53.

———. *Reflections on Gender and Science.* New Haven, Conn.: Yale University Press, 1985.

———. *Secrets of Life, Secrets of Death: Essays on Language, Gender and Science.* New York: Routledge, 1992.

Kerber, Linda K. " 'Ourselves and Our Daughters Forever': Women and the Constitution, 1787–1876." In *One Woman, One Vote: Rediscovering the Woman Suffrage Movement,* edited by Marjorie Spruill Wheeler, 21–36. Troutdale, Ore.: NewSage, 1995.

Kerin, Jacinta. "The Matter at Hand: Butler, Ontology and the Natural Sciences." *Australian Feminist Studies* 14, no. 29 (1999): 91–104.

Kevles, Bettyann Holtzmann. *Naked to the Bone: Medical Imaging in the Twentieth Century.* New Brunswick, N.J.: Rutgers University Press, 1997.

Kevles, Daniel J. *The Physicists: The History of a Scientific Community.* Cambridge, Mass.: Harvard University Press, 1995.

———. "The Physics, Mathematics, and Chemistry Communities: A Comparative Analysis." In *The Organization of Knowledge in America, 1860–1920,* edited by Alexandra Oleson and John Voss. Baltimore: Johns Hopkins University Press, 1979.

Kimball, Bruce A. *The "True Professional Ideal" in America: A History.* Cambridge, U.K.: Blackwell, 1992.

Kimmel, Michael. *Manhood in America: A Cultural History.* New York: Free Press, 1996.

King, William Harvey. *Electricity in Medicine and Surgery, Including the X-Ray.* New York: Boericke and Runyon, 1901.

Kingsley, Charles. "The Study of Physical Science: A Lecture to Young Men." *Popular Science Monthly* 1 (August 1872): 451–57.

Kline, Ronald. "Construing 'Technology' As 'Applied Science': Public Rhetoric of Scientists and Engineers in the United States, 1880–1945." *Isis* 86 (1995): 194–221.

Knight, Nancy. "The New Light: X-Rays and Medical Futurism." In *Imagining Tomorrow: History, Technology, and the American Future,* edited by Joseph J. Corn, 10–34. Cambridge, Mass.: MIT Press, 1986.

Koestler, Arthur. *The Sleepwalkers: A History of Man's Changing Vision of the Universe.* New York: Macmillan, 1959.

Kohler, Robert E. "The Ph.D. Machine: Building on the Collegiate Basis." *Isis* 81 (1990): 638–62.

Kolle, F. S. "The Harmful Effects of the X-Ray." *Electrical Engineer* 23 (19 May 1897): 521.

Korten, David. *The Post-Corporate World: Life After Capitalism*. San Francisco: Berrett-Koehler and Kumarian, 1999.

Kotz, Liz. "The Body You Want: Liz Kotz Interviews Judith Butler." *Artforum* 31 (November 1992): 82–89.

Kraemer, David. *Responses to Suffering in Classical Rabbinic Literature*. New York: Oxford University Press, 1995.

Krafft-Ebing, Richard von. *Psychopathia Sexualis*, translated by Franklin S. Klaf. New York: Stein and Day, 1965.

Kraut, Alan M. *Silent Travelers: Germs, Genes, and the "Immigrant Menace."* New York: Basic Books, 1994.

Kuhn, Thomas S. *The Structure of Scientific Revolutions*. 2d ed. Chicago: University of Chicago Press, 1970 [1962].

Kump, Robert J. *G. Stanley Hall's Efforts to Implement the Humboldtian University Ideal at Clark University in Worcester, Massachusetts, USA*. Bern, Switzerland: Haupt, 1996.

"The Laboratory in Modern Science." *Science* 3 (15 February 1884): 172–74.

Laderman, Gary. *The Sacred Remains: American Attitudes Toward Death, 1799–1883*. New Haven, Conn.: Yale University Press, 1993.

LaFee, Scott. "Perish the Thought: A Life in Science Sometimes Becomes a Death, Too." *San Diego Union-Tribune*, 5 December 2003, pp. F1, F4.

Larson, Edward J. *Summer for the Gods: The Scopes Trial and America's Continuing Debate over Science and Religion*. New York: Basic Books, 1997.

Latour, Bruno. *Science in Action: How to Follow Scientists and Engineers Through Society*. Cambridge, Mass.: Harvard University Press, 1987.

Law, John, and Annemarie Mol, eds., *Complexities: Social Studies of Knowledge Practices*. Durham, N.C.: Duke University Press, 2002.

Layton, Edwin T. "Mirror-Image Twins: The Communities of Science and Technology in 19th Century America." *Technology and Culture* 18 (1977): 562–80.

Leaman, W. "The Scope and Spirit of Scientific Research." *Mercersburg Review* 20 (October 1873): 522–37.

Lears, T. J. Jackson. *No Place of Grace: Antimodernism and the Transformation of American Culture, 1880–1920*. New York: Pantheon, 1981.

Lederer, Susan E. "The Controversy over Animal Experimentation in America, 1880–1914." In *Vivisection in Historical Perspective*, edited by Nicholaas A. Rupke, 236–58. London: Croom Helm, 1987.

———. "Film Review: *Arrowsmith*." *Isis* 84 (December 1993): 771–72.

———. "Political Animals: The Shaping of Biomedical Research Literature in Twentieth-Century America." *Isis* 83 (March 1992): 61–79.

———. *Subjected to Science: Human Experimentation in American before the Second World War*. Baltimore: Johns Hopkins University Press, 1995.

"Legal Proof of the Discovery of the Pole." *Bench and Bar* 18 (July 1909): 87–91.

Lenz, William E. *The Poetics of the Antarctic: A Study in Nineteenth-Century Cultural Perceptions*. New York: Garland, 1995.

Leonard, Charles Lester. "The Protection of the Roentgenologist." *New York Medical Journal* 86 (16 November 1907): 917–20.

———. "The X-Ray 'Burn': Its Productions and Prevention. Has the X Ray Any Therapeutic Properties?" *New York Medical Journal* 68 (2 July 1898): 18–20.

Leonard, Elizabeth D. *Yankee Women: Gender Battles in the Civil War*. New York: Norton, 1994.

Lepore, Jill. *The Name of War: King Philip's War and the Origins of American Identity*. New York: Vintage, 1999.

Leslie, Sir John. *Narrative of Discovery and Adventure in the Polar Seas and Regions.* Edinburgh: Oliver and Boyd, 1845.

Letvin, Alice Owen. *Sacrifice in the Surrealist Novel: The Impact of Early Theories of Primitive Religion on the Depiction of Violence in Modern Fiction.* New York: Garland, 1990.

Levinas, Emmanuel. "Useless Suffering." In *The Provocation of Levinas: Re-thinking the Other,* edited by Robert Bernasconi and David Wood, translated by Richard Cohen. London: Routledge, 1988.

Levine, George. *Dying to Know: Scientific Epistemology and Narrative in Victorian England.* Chicago: University of Chicago Press, 2002.

Lewis, Grace. *With Love from Gracie: Sinclair Lewis, 1912–1925.* New York: Harcourt, Brace, 1955.

Lewis, Sinclair. *Arrowsmith.* New York: New American Library, 1980 [1925].

——. *From Main Street to Stockholm: Letters of Sinclair Lewis, 1919–1930,* edited by Harrison Smith. New York: Harcourt, Brace, 1952.

Lightman, Alan. "Spellbound by the Eternal Riddle, Scientists Revel in Their Captivity." *New York Times,* 11 November 2003, p. D4.

Lindberg, David C., and Ronald L. Numbers, eds. *The Cambridge History of Science.* Cambridge, U.K.: Cambridge University Press, 2003.

——, eds. *God and Nature: Historical Essays on the Encounter between Christianity and Science.* Berkeley: University of California Press, 1986.

Lindee, Susan. "The Scientific Romance: Purity, Self-Sacrifice, and Passion in Popular Biographies of Marie Curie." Paper presented at the annual meeting of the History of Science Society, Atlanta, 8 November 1996.

Linderman, Gerald F. *Embattled Courage: The Experience of Combat in the American Civil War.* New York: Free Press, 1987.

Lingeman, Richard. *Sinclair Lewis: Rebel from Main Street.* New York: Random House, 2002.

Livingstone, David N., D. G. Hart, and Mark A. Noll, eds. *Evangelicals and Science in Historical Perspective.* New York: Oxford University Press, 1999.

Locke, John. *Two Treatises of Government,* edited by Peter Laslett. Cambridge, U.K.: Cambridge University Press, 1988.

London, Jack. "A Thousand Deaths." *Black Cat* 4 (May 1899): 33–42.

"The Longing of Science to Remain Useless." *Current Opinion* 71 (July 1921): 89–90.

Loomis, Chauncy C. *Weird and Tragic Shores: The Story of Charles Francis Hall, Explorer.* New York: Knopf, 1971.

López, Ian Haney. "The Prerequisite Cases." In *Racial Classification and History,* edited by E. Nathaniel Gates, 127–60. New York: Garland, 1997.

——. "Racial Restrictions in the Law of Citizenship." In *Racial Classification and History,* edited by E. Nathaniel Gates, 109–25. New York: Garland, 1997.

Lowie, Robert H. "Science." In *Civilization in the United States: An Inquiry by Thirty Americans,* edited by Harold E. Stearns, 151–61. New York: Harcourt Brace, 1922.

Löwy, Ilana. "Immunology and Literature in the Early Twentieth Century: *Arrowsmith* and *The Doctor's Dilemma.*" *Medical History* 32 (1988): 314–32.

Luibhéid, Eithne. *Entry Denied: Controlling Sexuality at the Border.* Minneapolis: University of Minnesota Press, 2002.

Lyman, Henry M. *Artificial Anaesthesia and Anaesthetics.* New York: Wood, 1881.

Lynch, Lisa L. "*Arrowsmith* Goes Native: Medicine and Empire in Fiction and Film." *Mosaic* 33 (December 2000): 193–208.

Lynch, Michael. "Springs of Action or Vocabularies of Motive?" In *Wellsprings of Achievement: Cultural and Economic Dynamics in Early Modern England and Japan,* edited by Penelope Gouk, 94–113. Aldershot, U.K.: Variorium, 1995.

MacMillan, Donald B. *How Peary Reached the Pole*. Boston: Houghton Mifflin, 1934.

Macpherson, C. B. *The Political Theory of Possessive Individualism: Hobbes to Locke*. Oxford: Clarendon, 1962.

Macy, John. *Walter James Dodd: A Biographical Sketch*. Boston: Houghton Mifflin, 1918.

Mangan, J. A., and James Walvin, eds. *Manliness and Morality: Middle-Class Masculinity in Britain and America, 1800–1940*. New York: St. Martin's, 1987.

Margon, Arthur. "Changing Models of Heroism in Popular American Novels, 1880–1920." *American Studies* 17 (fall 1976): 71–86.

"Martyrs of Science." *Literary Digest* 97 (2 June 1928): 19–20.

"Martyrs of Science." *Popular Mechanics* 44 (July 1925): 51.

Marx, Karl. *Early Writings*, translated by Rodney Livingstone and Gregor Benton. London: Penguin, 1992 [1844].

Marx, Leo. "The Idea of 'Technology' and Postmodern Pessimism." In *Does Technology Drive History? The Dilemma of Technological Determinism*, edited by Merritt Roe Smith and Leo Marx, 237–57. Cambridge, Mass.: MIT Press, 1994.

———. *The Machine in the Garden: Technology and the Pastoral Ideal in America*. New York: Oxford University Press, 1964.

Mather, Cotton. *A Brief Essay Upon the Cross*. Boston: Allen, 1714.

———. *The Sacrificer: An Essay Upon the Sacrifices Wherewith a Christian, laying a Claim to an Holy Priesthood, Endeavours to Glorify God*. Boston: Fleet, 1714.

Mauss, Marcel. *The Gift: The Form and Reason for Exchange in Archaic Societies*, translated by W. D. Hall. New York: Norton, 1990 [1924].

May, Henry F. *Protestant Churches and Industrial America*. New York: Octagon, 1977 [1949].

Maynard, Steven. "Rough Work and Rugged Men: The Social Construction of Masculinity in Working-Class History." *Labour/Le Travail* 23 (spring 1989): 159–69.

McKanan, Dan. *Identifying the Image of God: Radical Christians and Nonviolent Power in the Antebellum United States*. New York: Oxford University Press, 2002.

McPherson, James M. *For Cause and Comrades: Why Men Fought in the Civil War*. New York: Oxford University Press, 1997.

Meadowcroft, William Henry. *The ABC of X Rays*. New York: Excelsior, 1896.

Mendenhall, T. C. "The Relations of Men of Science to the General Public." *Proceedings of the American Association for the Advancement of Science* 39 (1890): 1–15.

Merback, Mitchell B. *The Thief, the Cross, and the Wheel: Pain and the Spectacle of Punishment in Medieval and Renaissance Europe*. London: Reaktion, 1999.

Merton, Robert K. *Science, Technology and Society in Seventeenth-Century England*. New York: Fertig, 1970.

Mialet, Hélène. "Do Angels Have Bodies? Two Stories about Subjectivity in Science." *Social Studies of Science* 29 (August 1999): 551–81.

Miller, John David. "H. A. Rowland and his Electromagnetic Researches." Ph.D. diss., Oregon State University, 1970.

Miller, Perry. *Errand into the Wilderness*. Cambridge, Mass.: Harvard University Press, 1956.

Miller, Perry, and Thomas H. Johnson. *The Puritans*. Vol. 1. New York: Harper and Row, 1963 [1938].

Miller, Stuart Creighton. *The Unwelcome Immigrant: The American Image of the Chinese, 1785–1882*. Berkeley: University of California Press, 1969.

Mills, C. Wright. "Situated Actions and Vocabularies of Motive." In *Power, People, and Politics*, 439–68. New York: Oxford University Press, 1963.

Mills, Charles K. *Mental Over-Work and Premature Disease among Public and Professional Men*. Washington, D.C.: Smithsonian Institution, 1885.

Mills, Charles W. "Black Trash." In *Faces of Environmental Racism*, edited by Laura Westra and Bill E. Lawson, 73–91. Boulder, Colo.: Rowman and Littlefield, 2001.

———. *The Racial Contract*. Ithaca, N.Y.: Cornell University Press, 1999.

Milne, A. A. *Winnie-the-Pooh*. New York: Dell, 1978 [1926].

Mirsky, Jeannette. *To the North! The Story of Arctic Exploration from Earliest Times to the Present*. New York: Viking, 1934.

Mitchell, S. Weir. "The Birth and Death of Pain." *Boston Medical and Surgical Journal* 135 (15 October 1896): 386.

Mizruchi, Susan L. *The Science of Sacrifice: American Literature and Modern Social Theory*. Princeton, N.J.: Princeton University Press, 1998.

Monell, Samuel Howard. *A System of Instruction in X-Ray Methods and Medical Uses of Light, Hot-Air, Vibration, and High Frequency Currents*. New York: Pelton, 1902.

Montrose, Louis. "The Work of Gender in the Discourse of Discovery." *Representations* 33 (winter 1991): 1–41.

Moore, Frank. *Women of the War: Their Heroism and Self-Sacrifice*. Hartford, Conn.: Scranton, 1866.

Moore, James R. *The Post-Darwinian Controversies: A Study of the Protestant Struggle to Come to Terms with Darwin in Great Britain and America, 1870–1900*. Cambridge, U.K.: Cambridge University Press, 1979.

"A Moral or Two from the Polar Controversy." *Century* 79 (March 1910): 793–94.

Morgan, Lewis Henry. *Ancient Society; or, Researches in the Lines of Human Progress from Savagery through Barbarism to Civilization*, edited by Eleanor Burke Leacock. Gloucester, Mass.: Smith, 1974 [1877].

Morris, David B. *The Culture of Pain*. Berkeley: University of California Press, 1991.

———. "An Invisible History of Pain: Early 19th-Century Britain and America." *Clinical Journal of Pain* 14, no. 3 (1998): 191–96.

Morris, G. S. "University Education." In *University of Michigan Philosophical Papers*, 1st ser., no. 1, pp. 1–36. Ann Arbor, Mich.: Andrews and Witherby, 1886.

Morrison, Toni. *Beloved*. New York: Signet, 1991 [1987].

Nansen, Fridtjof. *In Northern Mists: Arctic Exploration in Early Times*. Vol. 1. New York: Stokes, 1911.

Nash, Roderick. *Wilderness and the American Mind*. New Haven, Conn.: Yale University Press, 1967.

Neufeldt, Leonard N. "The Science of Power: Emerson's Views on Science and Technology in America." *Journal of the History of Ideas* 38 (April–June 1977): 329–44.

Newcomb, Raymond Lee, ed. *Our Lost Explorers: The Narrative of the Jeannette Arctic Expedition, as Related by the Survivors, and in the Records and Last Journals of Lieutenant De Long*. Hartford, Conn.: American Publishing Company, 1882.

Newcomb, Simon. "The Evolution of the Scientific Investigator." *Science* 20 (23 September 1904): 385–95.

Newman, Louise, ed. *Men's Ideas/Women's Realities: Popular Science, 1870–1915*. New York: Pergamon, 1985.

Nietzsche, Friedrich. *Daybreak: Thoughts on the Prejudices of Morality*, translated by R. J. Hollingdale. Cambridge, U.K.: Cambridge University Press, 1982 [1881].

———. *The Gay Science*, translated by Walter Kaufmann. New York: Vintage, 1974 [1882].

———. *On the Genealogy of Morals*, translated by Walter Kaufmann and R. J. Hollingdale. New York: Vintage, 1967 [1887].

Noble, David F. *A World Without Women: The Christian Clerical Culture of Western Science*. New York: Oxford University Press, 1992.

"The North Pole." *Century* 79 (November 1909): 152.

Norton, Charles Eliot. "The Advantages of Defeat." *Atlantic Monthly* 8 (September 1861): 360–65.

Novick, Peter. *That Noble Dream: The "Objectivity Question" and the American Historical Profession.* Cambridge, U.K.: Cambridge University Press, 1988.

Noyes, John K. *The Mastery of Submission: Inventions of Masochism.* Ithaca, N.Y.: Cornell University Press, 1997.

Noyes, William Albert. "Ira Remsen." *Science* 66 (16 September 1927): 243–46.

Noyes, William Albert, and James Flack Norris. *Biographical Memoir of Ira Remsen, 1846–1927.* Washington, D.C.: National Academy of Sciences, 1931.

Oakes, James. *Slavery and Freedom: An Interpretation of the Old South.* New York: Knopf, 1990.

Oleson, Alexandra, and John Voss, eds. *The Organization of Knowledge in Modern America, 1860–1920.* Baltimore: Johns Hopkins University Press, 1979.

"Operated on 72 Times: Roentgenologist Has Lost Eight Fingers and an Eye for Science." *New York Times,* 12 March 1926, p. 22.

Oreskes, Naomi. "Objectivity of Heroism? On the Invisibility of Women in Science." *Osiris* 11 (1996): 87–113.

"Our Great Debt to Science." *Scientific American* 77 (4 October 1890): 217.

Outram, Dorinda. "On Being Perseus: Travel and Truth in the Enlightenment." In *Geography and the Enlightenment,* edited by David N. Livingstone and Charles W. J. Withers, 281–94. Chicago: University of Chicago Press, 1999.

Owens, Larry. "Pure and Sound Government: Laboratories, Playing Fields, and Gymnasia in the Nineteenth-Century Search for Order." *Isis* 76 (1985): 182–94.

Owens, R. B. "Effect of Röntgen Rays on the Tissues." *Electrical World* 28 (19 December 1896): 759.

Palmquist, Peter E. *Elizabeth Fleischmann: Pioneer X-Ray Photographer.* Berkeley, Calif.: Magnes Museum, 1990.

Panchasi, Roxanne. "Reconstructions: Prosthetics and the Rehabilitation of the Male Body in World War I France." *differences* 7 (1995): 110–40.

Pateman, Carol. "Self-Ownership and Property in the Person: Democratization and a Tale of Two Concepts." *Journal of Political Philosophy* 10, no. 1 (2002): 20–53.

———. *The Sexual Contract.* Stanford, Calif.: Stanford University Press, 1988.

Patterson, Orlando. *Rituals of Blood: Consequences of Slavery in Two American Centuries.* Washington, D.C.: Civitas/Counterpoint, 1998.

Patterson, Thomas C. *Inventing Western Civilization.* New York: Monthly Review Press, 1997.

Paulsen, Friedrich. *The German Universities and University Study,* translated by Frank Thilly and William E. Elwang. New York: Scribner, 1906.

Pauly, Philip J. *Controlling Life: Jacques Loeb and the Engineering Ideal in Biology.* New York: Oxford University Press, 1987.

Pearson, Willie, Jr., and H. Kenneth Bechtel, eds. *Blacks, Science, and American Education.* New Brunswick, N.J.: Rutgers University Press, 1989.

Peary, Josephine Diebitsch. *My Arctic Journal: A Year Among Ice-Fields and Eskimos.* New York: Contemporary, 1893.

"Peary Likely to be a Rich Man." *Boston Sunday Globe,* 23 February 1908, p. 12.

Peary, Robert E. *Nearest the Pole.* London: Hutchinson, 1907.

Peirce, Charles Sanders. *The Essential Peirce: Selected Philosophical Writings,* edited by Nathan Houser et al., 2 vols. Bloomington: Indiana University Press, 1998.

Perkins, Judith. *The Suffering Self: Pain and Narrative Representation in the Early Christian Era.* London: Routledge, 1995.

Pernick, Martin S. *A Calculus of Suffering: Pain, Professionalism, and Anesthesia in Nineteenth-Century America*. New York: Columbia University Press, 1985.

Perrine, J. O. "The Fun of Being a Scientist," *Scientific Monthly* 27 (July 1928): 28–32.

Perry, Richard. *The Jeannette, and a Complete and Authentic Narrative Encyclopedia to the North Polar Regions*. San Francisco: Roman, 1883.

Petchesky, Rosalind Pollack. "The Body As Property: A Feminist Re-vision." In *Conceiving the New World Order: The Global Politics of Reproduction*, edited by Faye D. Ginsberg and Rayna Rapp, 387–406. Berkeley: University of California Press, 1995.

Pfahler, George E. "Fifty Years of Trials and Tribulations in Radiology." In *The American Roentgen Ray Society, 1900–1950: Commemorating the Golden Anniversary of the Society*. Springfield, Ill: Thomas, 1950.

Pickstone, John V. *Ways of Knowing: A New History of Science, Technology, and Medicine*. Manchester, U.K.: Manchester University Press, 2000.

Pink, Thomas, and M.W.F. Stone, eds. *The Will and Human Action: From Antiquity to the Present Day*. London: Routledge, 2004.

Pleck, Elizabeth H., and Joseph H. Pleck, eds. *The American Man*. Englewood Cliffs, N.J.: Prentice Hall, 1980.

Poovey, Mary. "Speaking of the Body: Mid-Victorian Constructions of Female Desire." In *Body/Politics: Women and the Discourses of Science*, edited by Mary Jacobus, Evelyn Fox Keller, and Sally Shuttlesworth, 24–46. New York: Routledge, 1990.

Porter, Charles Allen. "The Pathology and Surgical Treatment of Chronic X-Ray Dermatitis." *Transactions of the American Roentgen Ray Society* 9 (1908): 101–70.

Porter, Roy. *Flesh in the Age of Reason*. New York: Norton, 2003.

———. "Pain and Suffering." In *Companion Encyclopedia of the History of Medicine*, edited by W. F. Bynum and Roy Porter, vol. 2, pp. 1574–91. London: Routledge, 1993.

Porter, Theodore M. *Trust in Numbers: The Pursuit of Objectivity in Science and Public Life*. Princeton, N.J.: Princeton University Press, 1995.

"Presentation of the Rumford Medals to Professor Rowland." *Science* 3 (29 February 1884): 256–58.

"Professor Penck on the Polar Controversy." *Popular Mechanics* (December 1909): 786–87.

Pupin, Michael. *From Immigrant to Inventor*. New York: Scribner, 1923.

Pycior, Helena M., Nancy G. Slack, and Pnina G. Abir-Am, eds. *Creative Couples in the Sciences*. New Brunswick, N.J.: Rutgers University Press, 1996.

Rabinbach, Anson. *The Human Motor: Energy, Fatigue, and the Origins of Modernity*. Berkeley: University of California, 1990.

Rabinow, Paul. *Anthopos Today: Reflections on Modern Equipment*. Princeton, N.J.: Princeton University Press, 2003.

———. *Essays on the Anthropology of Reason*. Princeton, N.J.: Princeton University Press, 1996.

Radin, Margaret Jean. *Reinterpreting Property*. Chicago: University of Chicago Press, 1993.

Rao, Radhika. "Property, Privacy, and the Human Body." *Boston University Law Review* 80 (April 2000): 359–460.

Rawlins, Dennis. *Peary at the North Pole: Fact or Fiction?* Washington, D.C.: Luce, 1973.

Reed, Walter. "The Propagation of Yellow Fever: Observations Based on Recent Researches." *Medical Record* 60 (10 August 1901): 201–9.

Register, Woody. "Everyday Peter Pans: Work, Manhood, and Consumption in Urban America, 1900–1930." In *Boys and Their Toys? Masculinity, Class, and Technology in America*, edited by Roger Horowitz, 199–228. New York: Routledge, 2001.

Reik, Theodor. *Masochism in Modern Man*, translated by Margaret H. Beigel and Gertrud M. Kurth. New York: Farrar, Straus, 1941.

Reingold, Nathan. "Definitions and Speculations: The Professionalization of Science in America in the Nineteenth Century." In *The Pursuit of Knowledge in the Early American Republic*, edited by Alexandra Oleson and Sanborn C. Brown, 33–69. Baltimore: Johns Hopkins University Press, 1976.

Reiser, Stanley Joel. *Medicine and the Reign of Technology.* Cambridge, U.K.: Cambridge University Press, 1978.

Remarque, Erich Maria. *All Quiet on the Western Front.* New York: Ballantine, 1989 [1928].

"Resolutions of the International Geodetic Commission in Relation to the Unification of Longitudes and of Time." *Science* 2 (28 December 1883): 814–15.

Reuben, Julie A. *The Making of the Modern University: Intellectual Transformation and the Marginalization of Morality.* Chicago: University of Chicago Press, 1996.

Reverby, Susan M., ed. *Tuskegee's Truths: Rethinking the Tuskegee Syphilis Study.* Chapel Hill: University of North Carolina Press, 2000.

Rey, Roselyne. *The History of Pain,* translated by Louise Elliott Wallace, J. A. Cadden, and S. W. Cadden. Cambridge, Mass.: Harvard University Press, 1995.

Reynolds, Lawrence. "The History of the Use of the Roentgen Ray in Warfare." In *Classic Descriptions in Diagnostic Roentgenology,* edited by André J. Bruwer, vol. 2, 1307–17. Springfield, Ill: Thomas, 1964.

Rich, Adrienne. *The Dream of a Common Language, Poems 1974–1977.* New York: Norton, 1978.

Richards, Ellen. "The Elevation of Applied Science to an Equal Rank with the So-Called Learned Professions." In *Technology and Industrial Efficiency: A Series of Papers Presented at the Congress of Technology, Opened in Boston, Mass. April 10, 1911,* 124–28. New York: McGraw-Hill, 1911.

Richardson, Lyon N. "*Arrowsmith*: Genesis, Development, Variations." In *Sinclair Lewis's "Arrowsmith,"* edited by Harold Bloom, 37–47. New York: Chelsea House, 1988.

Ringrose, Marjorie, and Adam J. Lerner, eds. *Reimagining the Nation.* Buckingham, U.K.: Open University Press, 1993.

Rivers, W.H.R., and Henry Head. "A Human Experiment in Nerve Division." *Brain* 31 (November 1908): 323–450.

Robinson, Bradley. *Dark Companion.* New York: McBride, 1947.

Robinson, Michael F. *The Coldest Crucible: Arctic Exploration and American Culture, 1850–1910.* Chicago: University of Chicago Press, forthcoming.

Roediger, David R. *The Wages of Whiteness: Race and the Making of the American Working Class.* London: Verso, 1991.

Rosenberg, Charles E. *No Other Gods: On Science and American Social Thought.* Baltimore: Johns Hopkins University Press, 1997 [1976].

Rosenberg, Emily S. "Rescuing Women and Children." *Journal of American History* 89 (September 2002): 456–65.

Ross, Sydney. "*Scientist*: The Story of a Word." *Annals of Science* 18 (June 1962): 65–85.

Ross, W. Gillies. "Nineteenth-Century Exploration of the Arctic." In *North American Exploration,* edited by John Logan Allen, vol. 3, pp. 244–331. Lincoln: University of Nebraska Press, 1997.

Rossiter, Margaret W. *Women Scientists in America: Struggles and Strategies to 1940.* Baltimore: Johns Hopkins University Press, 1982.

Rotundo, E. Anthony. *American Manhood: Transformations in Masculinity from the Revolution to the Modern Era.* New York: Basic Books, 1993.

Rowbotham, Judith. "'Soldiers of Christ'? Images of Female Missionaries in Late Nineteenth-Century Britain: Issues of Heroism and Martyrdom." *Gender and History* 12 (April 2000): 82–106.

Rowland, Henry Augustus. "The Highest Aim of the Physicist." *American Journal of Science and Arts* 8 (1899): 401–11.

———. *The Physical Papers of Henry Augustus Rowland*. Baltimore: Johns Hopkins Press, 1902.

Rowlands, Michael. "Memory, Sacrifice, and the Nation." *New Formations* 30 (winter 1996): 8–17.

Russett, Cynthia Eagle. *Darwin in America: The Intellectual Response, 1865–1912*. San Francisco: Freeman, 1976.

———. *Sexual Science: The Victorian Construction of Womanhood*. Cambridge, Mass.: Harvard University Press, 1989.

Sánchez-Eppler, Karen. *Touching Liberty: Abolition, Feminism, and the Politics of the Body*. Berkeley: University of California Press, 1993.

Sandow, Eugen. *Body-Building, or Man in the Making: How to Become Healthy and Strong*. London: Gale and Polden, 1904.

Sarton, George. "The Discovery of the X-Rays." *Isis* 26 (1937): 340–69.

Savitt, Todd. "The Use of Blacks for Medical Experimentation and Demonstration in the Old South." *Journal of Southern History* 48 (August 1982): 331–48.

Savran, David. "The Sadomasochist in the Closet: White Masculinity and the Culture of Victimization." *differences* 8 (1996): 127–52.

Scarry, Elaine. *The Body in Pain: The Making and Unmaking of the World*. New York: Oxford University Press, 1985.

———. "Consent and the Body: Injury, Departure, and Desire." *New Literary History* 21 (autumn 1990): 867–96.

———. "The Merging of Bodies and Artifacts in the Social Contract." In *Culture on the Brink: Ideologies of Technology*, edited by Gretchen Bender and Timothy Druckrey, 85–97. Seattle: Bay, 1994.

Schaffer, Simon. "Astronomers Mark Time." *Science in Context* 2, no. 1 (1988): 115–45.

———. "Genius." In *Romanticism and the Sciences*, edited by Andrew Cunningham and Nicholas Jardine, 82–98. Cambridge, U.K.: Cambridge University Press, 1990.

Schiebinger, Londa. *Has Feminism Changed Science?* Cambridge, Mass.: Harvard University Press, 1999.

———. *The Mind Has No Sex? Women in the Origins of Modern Science*. Cambridge, Mass.: Harvard University Press, 1989.

Scholnick, Robert J., ed. *American Literature and Science*. Lexington: University of Kentucky Press, 1992.

Schorer, Mark. "Afterword." In *Arrowsmith*, by Sinclair Lewis. New York: New American Library, 1980.

———. *Sinclair Lewis: An American Life*. New York: McGraw-Hill, 1961.

———. *Sinclair Lewis: A Collection of Critical Essays*. Englewood Cliffs, N.J.: Prentice Hall, 1962.

Schweber, S. S. "Scientists As Intellectuals: The Early Victorians." In *Victorian Science and Victorian Values: Literary Perspectives*, edited by James Paradis and Thomas Postlewait, 1–37. New Brunswick, N.J.: Rutgers University Press, 1985.

Scott, Joan W. "'Experience.'" In *Feminists Theorize the Political*, edited by Judith Butler and Joan W. Scott, 22–40. New York: Routledge, 1992.

———. "Response to Gordon." *Signs* 15, no. 4 (1990): 859–60.

"Self-Sacrifice for the Sake of Science." *Current Literature* 31 (October 1901): 385–87.

Seltzer, Mark. *Bodies and Machines*. New York: Routledge, 1992.

Serlin, David. "Crippling Masculinity: Queerness and Disability in U.S. Military Culture, 1800–1945," *GLQ* 9, nos. 1 and 2 (2003): 149–79.

Serres, Michael. *The Natural Contract*, translated by Elizabeth MacArthur and William Paulson. Ann Arbor: University of Michigan Press, 1995.

Serwer, Daniel Paul. "The Rise of Radiation Protection: Science, Medicine and Technology in Society, 1896–1935." Ph.D. diss., Princeton University, 1977.

Seton, E. T. *Boy Scouts of America: A Handbook of Woodcraft, Scouting and Lifecraft*. New York: Doubleday, Page, 1910.

Shah, Nayan. *Contagious Divides: Epidemics and Race in San Francisco's Chinatown*. Berkeley: University of California Press, 2001.

Shapin, Steven. "'The Mind Is Its Own Place': Science and Solitude in Seventeenth-Century England," *Science in Context* 4, no. 1 (1990): 191–218.

———. *A Social History of Truth: Civility and Science in Seventeenth-Century England*. Chicago: University of Chicago, 1994.

Shapin, Steven, and Christopher Lawrence, eds., *Science Incarnate: Historical Embodiments of Natural Knowledge*. Chicago: University of Chicago Press, 1998.

Shapin, Steven, and Simon Schaffer. *Leviathan and the Air-Pump: Hobbes, Boyle, and the Experimental Life*. Princeton, N.J.: Princeton University Press, 1985.

Sharpe, Rebecca. "Random Remarks of a Lady Scientist." *Popular Science Monthly* 58 (1901): 548–50.

Sharpsteen, S. H. "The History of an X-Ray Burn." *Electrical Engineer* 24 (8 July 1897): 10–11.

Shelley, Mary Wollstonecraft. *Frankenstein, or, The Modern Prometheus*. Chicago: University of Chicago Press, 1982 [1818].

Shyrock, Richard H. "American Indifference to Basic Research During the Nineteenth Century." *Archives internationales d'histoire des sciences* 28 (1948): 50–65.

Siegal, Reva B. "Home As Work: The First Women's Rights Claims Concerning Wives' Household Labor, 1850–1880." *Yale Law Journal* 103 (March 1994): 1073–1217.

Siegel, Carol. *Male Masochism: Modern Revisions of the Story of Love*. Bloomington: Indiana University Press, 1995.

Silverman, Kaja. *Male Subjectivity at the Margins*. New York: Routledge, 1992.

Simmel, Georg. "A Chapter in the Philosophy of Value." *American Journal of Sociology* 5 (March 1900): 577–603.

———. *On Individuality and Social Forms: Selected Writings*, edited by Donald N. Levine. Chicago: University of Chicago Press, 1971.

"Sir John Franklin and the Arctic Regions." *North American Review* 71 (July 1850): 168–85.

[S.J.R.]. "Some Effects of the X-Rays on the Hands." *Nature* 54 (29 October 1896): 621.

Slosson, Edwin Emory. "The Relative Value of Life and Learning." *Independent*, 12 December 1895, p. 7.

Smith, Timothy L. *Revivalism and Social Reform: American Protestantism on the Eve of the Civil War*. New York: Harper and Row, 1957.

Smith, William Robertson. *Lectures on the Religion of the Semites*. New Brunswick, N.J.: Transaction, 2002 [1889].

Smucker, Samuel M. *Arctic Explorations and Discoveries during the Nineteenth Century*. New York: Lovell, 1886.

"Society Proceedings: Third Pan-American Medical Congress Held at Havana, Cuba, February 4–7, 1901." *Medical News* 78 (16 February 1901): 279–84.

Sontag, Susan. *Against Interpretation and Other Essays*. New York: Octagon, 1978.

Sosman, Merrill C. "Roentgenology at Harvard." *Harvard Medical Alumni Bulletin* 21 (April 1947): 65–76.

Spencer, Frank ed. *Piltdown Papers: 1908–1955*. London: Oxford University Press, 1990.

Spillers, Hortense J. "Mama's Baby, Papa's Maybe: An American Grammar Book." *diacritics* 17 (summer 1987): 65–81.

Stanley, Amy Dru. "Conjugal Bonds and Wage Labor: Rights of Contract in the Age of Emancipation." *Journal of American History* 75 (September 1988): 471–500.

——. *From Bondage to Contract: Wage Labor, Marriage, and the Market in the Age of Slave Emancipation.* New York: Cambridge University Press, 1998.

——. "'The Right to Possess All the Faculties That God Has Given': Possessive Individualism, Slave Women, and Abolitionist Thought." In *Moral Problems in American Life: New Perspectives on Cultural History,* edited by Karen Halttunen and Lewis Perry, 123–43. Ithaca, N.Y.: Cornell University Press, 1998.

Stead, W. T. "Character Sketch and Interview, Dr. F. A. Cook." *Review of Reviews* 40 (October 1909): 323–39.

Stepan, Nancy. *The Idea of Race in Science: Great Britain, 1800–1960.* Hamden, Conn.: Archon, 1982.

——. "The Interplay between Socio-Economic Factors and Medical Science: Yellow Fever Research, Cuba and the United States." *Social Studies of Science* 8 (November 1978): 397–423.

Sternberg, George M. "Transmission of Yellow Fever by the Mosquito." *Journal of Social Science* 39 (1 November 1901): 84–99.

Stine, W. M. "Effect on the Skin of Exposure to the Roentgen Tubes." *Electrical Review* 29 (18 November 1896): 250.

——. "Physiological Effects of the Röntgen Tube." *Electrical World* 28 (19 December 1896): 748.

Stocking, George W., Jr. *Race, Culture, and Evolution.* New York: Free Press, 1968.

Stone, Christopher D. *Should Trees Have Standing? Toward Legal Rights for Natural Objects.* Palo Alto, Calif.: Tioga, 1988.

Stone, Ian R. "The Franklin Expedition in Parliament." *Polar Record* 32 (1996): 209–16.

Stone-Mediatore, Shari. "Chandra Mohanty and the Revaluing of 'Experience.'" In *Decentering the Center: Philosophy for a Multicultural, Postcolonial, and Feminist World,* edited by Uma Narayan and Sandra Harding, 110–27. Bloomington: Indiana University Press, 2000.

Stout, Arthur B. *Chinese Immigration and the Physiological Causes of the Decay of a Nation.* San Francisco: Agnew and Deffelbach, 1862.

Strickland, Stuart Walker. "The Ideology of Self-Knowledge and the Practice of Self-Experimentation." *Eighteenth-Century Studies* 31, no. 4 (1998): 453–71.

Summer, Francis B. "Some Perils Which Confront Us As Scientists." *Scientific Monthly* 8 (March 1919): 258–74.

Sumner, William Graham. *What Social Classes Owe to Each Other.* New York: Harper, 1883.

Sweet, William Warren. *The Story of Religions in America.* New York: Harper, 1930.

Takaki, Ronald. *Iron Cages: Race and Culture in Nineteenth Century America.* New York: Knopf, 1979.

Tanner, Amy E. "History of Clark University Through the Interpretation of the Will of the Founder." Unpublished manuscript. Clark University Archives, Worcester, Mass., 1908.

Taylor, Charles. "The Person." In *The Category of the Person: Anthropology, Philosophy, History,* edited by Michael Carrithers, Steven Collins, and Steven Lukes, 257–81. Cambridge, U.K.: Cambridge University Press, 1985.

Terrall, Mary. "Gendered Spaces, Gendered Audiences: Inside and Outside the Paris Academy of Sciences." *Configurations* 2 (1995): 207–32.

———. "Heroic Narratives of Quest and Discovery." *Configurations* 6 (1998): 223–42.

Tesla, Nicola. "Tesla on the Hurtful Actions of the Lenard and Roentgen Tubes." *Electrical Review* 30 (5 May 1897): 207, 211.

Thomson, Elihu. "Roentgen Rays Act Strongly on the Tissues." *Electrical Engineer* 22 (25 November 1896): 534.

———. "Roentgen Rays Act Strongly on the Tissues." *Electrical Review* 29 (25 November 1896): 260.

———. "Röntgen Rays Act Strongly on the Tissues." *Electrical World* 28 (28 November 1896): 666.

———. "Some Notes on Roentgen Rays." *Electrical Engineer* 22 (18 November 1896): 520–21.

———. "Some Recent Röntgen-Ray Work." *Electrical World* 28 (10 October 1896): 415–16.

Tichi, Cecilia. *Shifting Gears: Technology, Literature, and Culture in Modernist America.* Chapel Hill: University of North Carolina Press, 1987.

Tissandier, Gaston. *Les martyrs de la science.* Paris: Dreyfous, 1880.

Tocqueville, Alexis de. *Democracy in America,* translated by Henry Reeve, vol. 2. New York: Vintage, 1945.

Townsend, Kim. *Manhood at Harvard: William James and Others.* Cambridge, Mass.: Harvard University Press, 1996.

Trachtenberg, Alan. *The Incorporation of America: Culture and Society in the Gilded Age.* New York: Hill and Wang, 1982.

Traweek, Sharon. *Beamtimes and Lifetimes: The World of High Energy Physicists.* Cambridge, Mass.: Harvard University Press, 1988.

Turner, James. *Reckoning with the Beast: Animals, Pain, and Humanity in the Victorian Mind.* Baltimore: Johns Hopkins University Press, 1980.

Turner, Victor. *Dramas, Fields, and Metaphors: Symbolic Action in Human Society.* Ithaca, N.Y.: Cornell University Press, 1974.

"Undergoes 50th Operation: Dr. Baetjer of Johns Hopkins is a Victim of X-Ray Infections." *New York Times,* 2 May 1924, p. 26.

"Untoward Effects of X-Rays." *Boston Medical and Surgical Journal* 152 (1905): 173–74.

Vanderbilt, Kermit. *Charles Eliot Norton: Apostle of Culture in a Democracy.* Cambridge, Mass.: Belknap, 1959.

Veysey, Laurence R. *The Emergence of the American University.* Chicago: University of Chicago Press, 1965.

Vivian, Thomas J. "John Chinaman in San Francisco." *Scribner's Monthly* 12 (1876): 862–72.

von Holst, Hermann E. "The Need of Universities in the United States." *Educational Review* 5 (1893): 105–19.

Wagner, Belinda J. "A Scientist in Search of Balance." *Sojourner* 22 (January 1997): 17–18.

Walker, W. H. "Chemical Research and Industrial Progress: What Commerce Owes to Chemistry." *Scientific American Supplement* 72 (1 July 1911): 14–16.

Walworth, Helen Harelin. "Field Work by Amateurs." *Science* 1 (16 October 1880): 198–99.

Warner, John Harley. "Remembering Paris: Memory and the American Disciples of French Medicine in the Nineteenth Century." *Bulletin of the History of Medicine* 65 (fall 1991): 301–25.

Warner, Margaret. "Hunting the Yellow Fever Germ: The Principle and Practice of Etiological Proof in Late Nineteenth-Century America." *Bulletin of the History of Medicine* 59 (1985): 361–82.

Weber, Max. *From Max Weber: Essays in Sociology,* edited and translated by H. H. Gerth and C. Wright Mills. New York: Oxford University Press, 1946.

———. *The Protestant Ethic and the Spirit of Capitalism*, translated by Talcott Parsons. New York: Scribner, 1958 [1904–5].

Weems, John Edward. *Peary: The Explorer and Man*. Boston: Houghton Mifflin, 1967.

———. *Race for the Pole*. New York: Holt, 1960.

Weir, Allison. *Sacrificial Logics: Feminist Theory and the Critique of Identity*. New York: Routledge, 1996.

Welter, Barbara. "The Cult of True Womanhood: 1820–1860." In *The American Family in Social-Historical Perspective*, edited by Michael Gordon, 224–50. New York: St. Martin's, 1973.

Weyprecht, Karl. "Scientific Work of the Second Austro-Hungarian Polar Expedition, 1872–4." *Royal Geographic Society Journal* 45 (1875): 19–33.

Wheeler, William Morton. "The Organization of Research." *Science* 53 (1921): 53–67.

[Whewell, William]. Review of *On the Connexion of the Physical Sciences*. *Quarterly Review* 51 (March and June 1834): 54–68.

White, Andrew Dickson. *A History of the Warfare of Science with Theology in Christendom*. Vol. I. New York: Appleton, 1919 [1896].

White, J. C. "Dermatitis Caused by X-Rays." *Boston Medical and Surgical Journal* 135 (3 December 1896): 583.

Whittemore, Gilbert F. "The National Committee on Radiation Protection, 1928–1960: From Professional Guidelines to Government Regulation." Ph.D. diss., Harvard University, 1986.

[W.H.W.]. "Lawyers' Fees." *Central Law Journal* 11 [1880]: 219–20.

Wiegman, Robyn. "Intimate Publics: Race, Property, and Personhood." *American Literature* 74 (December 2002): 859–85.

Wilkinson, Doug. *Arctic Fever: The Search for the Northwest Passage*. Toronto: Clarke, Irwin, 1971.

Williams, Patricia J. *The Alchemy of Race and Rights*. Cambridge, Mass.: Harvard University Press, 1991.

Williams, Raymond. *Culture and Society, 1780–1950*. New York: Harper and Row, 1958.

———. *Keywords*. New York: Oxford University Press, 1976.

Wilson, Daniel J. *Science, Community, and the Transformation of American Philosophy, 1860–1930*. Chicago: University of Chicago Press, 1990.

Wilson, Eric G. *The Spiritual History of Ice: Romanticism, Science, and the Imagination*. New York: Palgrave Macmillan, 2003.

Winter, Alison. *Mesmerized: Powers of Mind in Victorian Britain*. Chicago: University of Chicago Press, 1998.

Wolbach, S. Burt. "Summary of the Effects of Repeated Roentgen-Ray Exposures Upon the Human Skin, Antecedent to the Formation of Carcinoma." *American Journal of Roentgenology and Radiology* 13 (1925): 139–43.

Wolf, Hazel Catherine. *On Freedom's Altar: The Martyr Complex in the Abolition Movement*. Madison: University of Wisconsin Press, 1952.

Wolfe, Alan. "Anti-American Studies." *New Republic*, 10 February 2003, pp. 25–32.

"The Woman Who Takes the Best Radiographs in the World." *San Francisco Chronicle*, 3 June 1900, p. 30.

Wong, K. Scott. "Cultural Defenders and Brokers: Chinese Responses to the Anti-Chinese Movement." In *Claiming America: Constructing Chinese American Identities during the Exclusion Era*, edited by K. Scott Wong and Sucheng Chan, 3–40. Philadelphia: Temple University Press, 1998.

Woodbury, David O. *Elihu Thomson: Beloved Scientist.* Boston: Museum of Science, 1960.

Wråkberg, Urban. *Vetenskapens vikingatåg: Perspektiv på svensk polarforskning, 1860–1930.* Uppsala: Institutionen för Idé-och Lärdomshistoria, 1995.

Wright, Theon. *The Big Nail: The Story of the Cook-Peary Feud.* New York: Day, 1970.

Young, Iris Marion. "Impartiality and the Civic Public: Some Implications of Feminist Critique of Moral and Political Theory." In *Feminism As Critique,* edited by Seyla Benhabib and Drucilla Cornell, 57–76. Minneapolis: University of Minnesota, 1987.

INDEX